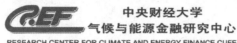

2017 中国气候融资报告

2017 China Climate Financing Report

刘倩 崔莹 著

责任编辑：肖　炜　董梦雅
责任校对：张志文
责任印制：张也男

图书在版编目（CIP）数据

2017 中国气候融资报告（2017 Zhongguo Qihou Rongzi Baogao）/刘倩，崔莹著．—北京：中国金融出版社，2018.6
ISBN 978 - 7 - 5049 - 9548 - 3

Ⅰ.①2… Ⅱ.①刘…②崔 Ⅲ.①气候变化—融资—研究报告—中国—2017 Ⅳ.①P467 - 05

中国版本图书馆 CIP 数据核字（2018）第 079781 号

出版 发行	中国金融出版社
社址	北京市丰台区益泽路 2 号
市场开发部	（010）63272190，66070804（传真）
网上书店	http://www.chinafph.com
	（010）63286832，63365686（传真）
读者服务部	（010）66070833，62568380
邮编	100071
经销	新华书店
印刷	保利达印务有限公司
尺寸	169 毫米 × 239 毫米
印张	10.5
字数	163 千
版次	2018 年 6 月第 1 版
印次	2018 年 6 月第 1 次印刷
定价	48.00 元

ISBN 978 - 7 - 5049 - 9548 - 3
如出现印装错误本社负责调换　联系电话(010)63263947

关于本报告

中央财经大学绿色金融国际研究院（IIGF）

　　中央财经大学绿色金融国际研究院（以下简称绿金院）是国内首家以推动绿色金融发展为目标的开放型、国际化的研究院。绿金院前身为中央财经大学气候与能源金融研究中心，是中国金融学会绿色金融专业委员会的常务理事单位。绿金院以营造富有绿色金融精神的经济环境和社会氛围为己任，致力于打造国内一流、世界领先的具有中国特色的专业化金融智库。

中央财经大学气候与能源金融研究中心（RCCEF）

　　中央财经大学气候与能源金融研究中心成立于2011年9月，已连续七年发布《中国气候融资报告》，基于广义的全球气候融资概念，形成了气候融资流的分析框架，构建了中国气候融资需求模型，并从国际气候资金治理以及中国气候融资的发展角度，逐年进行深入分析，积累了一系列的气候融资研究成果。基于长期信任与合作的基础，气候与能源金融研究中心与财政部建立了部委共建学术伙伴关系。

指导

　　王　遥　　　　中央财经大学绿色金融国际研究院院长、教授、博导
　　　　　　　　　中国金融学会绿色金融专业委员会副秘书长

作者

　　刘　倩　　　　中央财经大学气候与能源金融研究中心执行主任、教授
　　崔　莹　　　　中央财经大学气候与能源金融研究中心高级研究员
　　Mathias Lund Larsen　中央财经大学气候与能源金融研究中心研究员
　　许寅硕　　　　中央财经大学气候与能源金融研究中心主任助理

罗谭晓思	中央财经大学气候与能源金融研究中心研究员
姚颖志	中央财经大学气候与能源金融研究中心助理研究员
娄惠源	中央财经大学气候与能源金融研究中心研究助理
张东侯	中央财经大学气候与能源金融研究中心研究助理
徐　蕾	中央财经大学气候与能源金融研究中心研究助理

前　言

2015年签署的《巴黎协定》已进入具体的落实阶段。而过去这一年是"黑天鹅事件"频发的一年，全球政治经济格局发生了深刻的重组与变革，对气候治理格局与气候融资体系亦造成了一定程度的冲击，其中最具代表性的事件是美国总统特朗普于2017年6月1日宣布退出《巴黎协定》。尽管如此，各国共同发展绿色金融并携手应对气候变化的决心并未受到太大的动摇，尤其是新兴经济体近年来积极应对气候变化，在开展"南南合作"为其他发展中国家提供气候资金支持方面作出了较大的贡献，在这一系列行动中，中国的表现可谓脱颖而出。

中国近几年对生态文明建设、环境保护、低碳发展与应对气候变化高度重视，并将绿色金融上升为国家战略。在党的十九大报告中，习近平总书记提到"坚持节约资源和保护环境的基本国策，像对待生命一样对待生态环境"，"绿色"一词被反复提及，从党的十八大的一次提及到十九大的十五次之多，增幅之大足以彰显中国走可持续发展道路的坚定决心。中国在国际上也积极倡导绿色金融的发展，不仅是《巴黎协定》签署的促成者，也推动绿色金融纳入G20杭州峰会议程，并助力其他发展中国家应对气候变化的工作。显然，中国正逐渐以引领者的姿态出现在国际气候治理的舞台之上。

气候资金长期以来是国际气候谈判及应对气候变化行动能否成功的关键要素。目前全球的焦点集中在如何填补应对气候变化的资金供给和需求的巨大鸿沟上，中央财经大学气候与能源金融研究中心（RCCEF）从2011年开始紧密追踪国际气候融资的发展动态，时至今日已经七年。本报告在过去累积的经验与方法学之上，梳理了过去一年的最新动态，反思了现存的气候融资体系的问题，在对如何撬动私人资本、加强气候适应资金供给等方面做了分析，希望能为中国乃至国际社会贡献来自发展中国家的研究智慧，并为政策制定者建言献策。

英文缩写索引表

英文简写	英文全称	中文对应名称
A		
AAU	Assigned Amount Units	分配数量单位
ABS	Asset-Backed Security	资产支持证券
ACCF	Africa Climate Change Fund	非洲气候变化基金
ASAP	Adaptation for Smallholder Agriculture Programme	小型农业适应计划
ADB	Asian Development Bank	亚洲开发银行
AF	Adaptation Fund	适应基金
AfDB	African Development Bank	非洲开发银行
AIIB	Asian Infrastructure Investment Bank	亚洲基础设施投资银行
ARC	Africa Risk Capacity	非洲风险防范能力机制
B		
BFIs	Bilateral Financial Institutions	双边金融机构
C		
CBFF	Congo Basin Forest Fund	刚果盆地森林基金
CCCFs	County Climate Change Funds	县级气候变化基金
CCCG	Climate Change Coordination Group	气候变化协调小组
CCER	China Certified Emission Reduction	中国核证自愿减排量
CCF	Climate Change Fund	气候变化基金
CCRIF	Caribbean Catastrophe Risk Insurance Facility	加勒比巨灾风险保险基金
CCWG	Climate Change Working Group	气候变化工作组
CCXG	Climate Change Expert Group	气候变化专家组
CDKN	Climate and Development Knowledge Network	气候与发展知识网络
CDM	Clean Development Mechanism	清洁发展机制
CDMF	Clean Development Mechanism Fund	清洁发展机制基金
CER	Certification Emission Reduction	核证减排量
CFIF	Climate Finance Innovation Facility	气候融资创新机制

续表

英文简写	英文全称	中文对应名称
CIDA	Canadian International Development Agency	加拿大国际开发署
CIFs	Climate Investment Funds	气候投资基金
CTF	Clean Technology Fund	清洁技术基金
COP	Conferences of the Parties	联合国气候变化框架公约缔约方大会
CPI	Climate Policy Initiative	气候政策倡议
CRGE	Climate Resilient Green Economy	气候韧性绿色经济机制
D		
DFIs	Developmental Finance Institutions	发展性金融机构
DECC	Department of Energy and Climate Change	能源与气候变化司
DFAT	Department of Foreign Affairs and Trade	外交与贸易部
DFID	Department for International Development	国际发展部
E		
EBRD	European Bank for Reconstruction and Development	欧洲复兴开发银行
EC	European Commission	欧盟委员会
XCF	Extreme Climate Facility	极端天气防范机制
EIB	European Investment Bank	欧洲投资银行
ERU	Emission Reduction Unit	排放减量单位
F		
FAO	Food and Agriculture Organization	联合国粮食和农业组织
FCPF	The Forest Carbon Partnership Facility	森林碳伙伴基金
FDI	Foreign Direct Investment	外商直接投资
FFEM	French Global Environment Facility	法国全球环境基金
FIP	Forest Investment Program	森林投资项目
FoF	Fund of Funds	母基金
FTT	Financial Transaction Tax	金融交易税
G		
GCF	Green Climate Fund	绿色气候基金
GDP	Gross Domestic Product	国内生产总值
GEF	Global Environmental Facility	环球环境基金
GLO	Global Legislators Organization	全球立法组织

续表

英文简写	英文全称	中文对应名称
GIL	Global Innovation Lab	全球创新实验室
GIZ	Deutsche Gesellschaft für Internationale Zusammenarbeit	德国国际合作机构
GRICCE	Grantham Research Institute on Climate Change and the Environment	格兰瑟姆气候变化与环境研究所
I		
IADB	Inter-American Development Bank	泛美开发银行
IBRD	International Bank for Reconstruction and Development	国际复兴开发银行
ICI	International Climate Initiative	国际气候倡议
IDA	International Development Association	国际开发协会
IEA	International Energy Agency	国际能源署
IET	International Emission Trading	排放贸易机制
IFAD	International Fund for Agricultural Development	国际农业发展基金
IFC	International Finance Corporation	国际金融公司
INDC	Intended Nationally Determined Contributions	国家自主贡献
IPCC	Intergovernmental Panel on Climate Change	政府间气候变化委员会
J		
JFSF	Japan Fast Start Fund	日本快速启动基金
JI	Joint Implementation	联合履约机制
JICA	Japan International Cooperation Agency	日本国际合作署
K		
KfW	Kreditanstalt für Wiederaufbau	德国复兴信贷银行
L		
LDCF	Least Developed Countries Fund	最不发达国家基金
LDCs	Least Developed Countries	最不发达国家
M		
MDBs	Multilateral Development Banks	多边开发银行
MFIs	Multilateral Financial Institutions	多边金融机构
MIGA	Multinational Investment Guarantee Agency	多边投资担保机构
MP	Montreal Protocol	蒙特利尔议定书
MRV	Monitoring, Reporting and Verification	监测、报告和核证体系

续表

英文简写	英文全称	中文对应名称
N		
NAMAs	Nationally Appropriate Mitigation Actions	国家适当减缓行动
NAPAs	National Adaptation Plans of Action	国家适应行动计划
NAPs	National Adaptation Plans	国家适应计划
NCF	National Climate Fund	国家气候基金
NDBs	National Development Bank	国家开发银行
NORAD	Norwegian Agency for International Development Cooperation	挪威国际发展合作署
O		
ODA	Official Development Assistance	官方开发援助
ODI	Overseas Development Institute	海外发展机构
OE	Outbreak & Epidemic	传染性疾病爆发防范机制
OPIC	Overseas Private Investment Corporation	海外私人投资公司
OECD	Organization for Economic Co-operation and Development	经济合作与发展组织
P		
PPCR	Pilot Program for Climate Resilience	气候韧性试点项目
PMR	Partnership for Market Readiness	市场准备伙伴关系
PPP	Public-Private Partnership	政府与社会资本合作模式
R		
RC	Replica Coverage	复制性覆盖
REDD+	Reducing Emissions from Deforestation and Degradation	减少毁林和森林退化造成的温室气体排放
S		
SCCF	Special Climate Change Fund	气候变化特别基金
SCF	Strategic Climate Fund	战略气候基金
SGRCC	St George Regional Collaboration Center	圣乔治地区合作中心
SIDA	Swedish International Development Cooperation Agency	瑞典国际发展合作署
SPV	Special Purpose Vehicle	特殊目的实体
U		
UNDP	United Nations Development Programme	联合国开发计划署
UNEP	United Nations Environment Programme	联合国环境规划署
UNFCCC	United Nations Framework Convention on Climate Change	联合国气候变化框架公约

续表

英文简写	英文全称	中文对应名称
UNISDR	United Nations International Strategy for Disaster Reduction	联合国减灾署
UN–REDD	United Nations Reducing Emissions from Deforestation and Forest Degradation	减少发展中国家毁林和森林退化所致排放量联合国合作方案
USAID	United States Agency for International Development	美国国际开发署
W		
WB	World Bank	世界银行
WBG	World Bank Group	世界银行集团
WFF	Water Financing Facility	水资源融资机制
WMO	World Meteorological Organization	世界气象组织
WRI	the World Resource Institution	世界资源研究所

目 录
Contents

一、全球气候融资进展 ·· 1
 (一) 气候融资缺口仍在持续扩大 ································· 1
 (二) 全球气候公共物品尚未形成一体化的资金供给机制 ··········· 2
 (三) 资金机制之间的协同度有待提高 ····························· 4
 (四) 新旧多边发展机构需要寻求协同机制与模式创新 ············· 5

二、中国气候融资进展 ·· 10
 (一) 气候变化政策体系逐步完善 ································· 10
 (二) PPP 模式在绿色低碳领域的运用进一步深化 ················· 15
 (三) 绿色金融工具撬动气候资金能力不断显现 ···················· 19
 (四) 中国的绿色投资实践逐步走向国际 ·························· 21

三、其他发展中国家气候融资进程与合作 ····························· 28
 (一) 其他发展中国家气候融资取得的进展 ························ 28
 (二) 发展中国家积极开展气候援助与合作 ························ 33
 (三) 发展中国家的需求及中国对策 ······························ 33

四、特朗普政府对全球气候融资的影响 ································ 36
 (一) 全球气候资金流的影响 ····································· 37
 (二) 全球气候治理格局的变化 ··································· 39
 (三) 中国在全球气候融资中的新角色 ···························· 42

五、多边开发银行的合作 ·· 44
（一）多边开发银行间的合作潜力巨大 ······················· 44
（二）亚投行与欧投行的合作 ······································ 46

六、政策建议 ·· 52
（一）坚持履行《巴黎协定》，推动全球建立一体化的气候公共
物品供给机制 ·· 52
（二）推动新兴多边机构在应对气候变化领域的渠道作用和多方面
创新 ·· 53
（三）以气候变化南南合作为突破口，加强我国气候融资软实力 ········ 54

一、全球气候融资进展

（一）气候融资缺口仍在持续扩大

2015年12月12日在巴黎气候变化大会上通过的《巴黎协定》提出要把全球平均温度升幅较工业化前水平控制在2℃以内，并向1.5℃的目标努力。从最新的气候投融资情况来看，全球气候资金缺口正在持续扩大。根据国际能源署（International Energy Agency，IEA）的预测，为实现《巴黎协定》目标，2015~2030年仅能源领域就需要16.5万亿美元。另外，根据Sam Fankhauser等人的估算方法，未来10年额外新增的减缓和适应资金需求总额将达到6300亿美元/年，其中中国的气候资金需求将达2050亿美元/年[①]。根据联合国政府间气候变化专家小组（Intergovernmental Panel on Climate Change，IPCC）的统计，2010年到2050年期间发展中国家适应资金需求规模为700~1000亿美元/年。而据联合国环境规划署（United Nations Environment Program，UNEP）估算，到2030年，适应资金的年需求规模将达到1400亿~3000亿美元，到2050年，适应资金需求将达到2800亿~5000亿美元。麦肯锡测算，2015年至2030年可持续性基础设施的资金缺口达到39万~51万亿美元，其中中等收入国家的资金缺口占比达65%[②]，尽管各机构的预测口径不尽相同，但预测结果都显示全球气候资金缺口持续扩大。

根据《巴黎协定》提出的资金安排，在2020年以后发达国家向发展中国家每年至少动员1000亿美元的资金支持。对于发达国家的履约情况，根据经济合作与发展组织（Organization for Economic Cooperation and Development，

---

① Sam Fankhauser, Aditi Sahni, Annie Savvas & John Ward. Where are the gaps in climate finance? [J]. 2016, 8 (3). Climate and Development, 2016.

② McKinsey. 2016. Financing Change：How to mobilize private sector financing for sustainable infrastructure. 39万亿美元与51万亿美元的数据差异是根据不同计算方法得到的。39万亿美元为保守估计，剔除了中国在基础设施投资上13.4%的高增长，按照全球平均增长的1.8%计算，51万亿美元为较激进的估算方式，假设中国能保持历史增长率，则将全球平均增长率拉高至4.3%。

OECD）的报告，发达国家通过公共和私营部门向发展中国家动员的气候资金规模在2013年和2014年分别为522亿美元和618亿美元，其中公共资金分别为379亿美元和435亿美元①。这份报告的统计口径受到很多国家的质疑。2017年11月德国波恩气候大会上，发达国家就1000亿美元资金安排问题依然未达成一致。即使将私人资金和官方开发援助（Ofricial Development Assistance，ODA）纳入统计，且2020年发达国家承诺的1000亿美元资金能顺利到位，这些资金也远远不足以应对气候变化带来的巨大挑战。

（二）全球气候公共物品尚未形成一体化的资金供给机制

在过去的25年中，全球气候资金供给渠道不断扩充，气候融资体系日益朝着复杂的方向演化。早期的气候融资对官方发展援助体系具有很强的路径依赖，偏重"捐资国向受援国"的单向援助模式。金融危机后，发达国家ODA增长乏力，新兴市场国家与私人融资机制的影响力逐步增强，基于OECD援助体系的多边气候机制虽然能够惠及全球，但其贡献的气候资金量占比在所有渠道中最小，双边渠道是气候资金的主要来源，它以高效直接的独特优势受到资助国家的欢迎，资金贡献量呈上升趋势。近年来，绿色气候基金（Green Climate Fund，GCF）作为联合国气候变化框架公约（United Nations Framework Convention on Climate Change，UNFCCC）专属的融资机构得以建立，另外，非洲风险防范能力机制（Africa Risk Capability，ARC）与加勒比巨灾风险保险基金（Caribbean Catastrophe Risk Insurance Facility，CCRIF）的建立标志着适应领域的融资举措与风险管理机制取得了突破性进展（格局概览如图1-1所示）。然而，目前全球气候融资体系仍然存在较大的困局，如整体碎片化、协调度较低，难以体现新兴市场国家和其他发展中国家的需求与贡献，难以支撑全球公共物品供给的持续性与公平性。

国际公共气候资金主要来自能够带来减缓与适应效应的ODA。双边气候资金占ODA比重已经从2003年的4%上升到了2013~2014年的18%②。

---

① OECD, CPI. Climate Finance in 2013 - 14 and the USD 100 billion goal [EB/OL]. http://www.oecd-ilibrary.org/docserver/download/9715381e.pdf?expires=1517901239&id=id&accname=guest&checksum=75A4697E7FA9ADF080FE02BD7169CB4B.

② Tracy C, Jan K, Annaka P. Climate finance shadow report [R/OL]. [2017-06-10]. https://www.oxfam.org/sites/www.oxfam.org/files/file_attachments/bp-climate-finance-shadow-report-031116-en.pdf.

一、全球气候融资进展

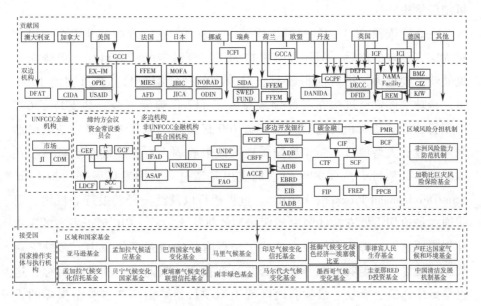

图 1-1　全球气候公共物品供给格局①

传统的国际援助体系是富裕国家通过 ODA 向贫困国家转移财政资金，意图建立"援助国—受援国"的单向关系，同时前者在后者的政治与社会等各个领域中也实施强大的影响力。因此，ODA 相当于援助国家财政部门公共事务在国际事务中的延伸。而应对气候变化资金机制则应属支持全球公共物品供给的全球公共财政机制，如何最大化资金在支持全球减缓与适应效果上的效益，在全球范围内实现资金利用的效率与公平是其核心目标②，因此二者在出发点、干预重点上存在区别。即使发展援助与气候融资在某些领域存在一定程度的协同效应，但以 ODA 作为气候资金机制的主要部分，仍然存在牺牲发展目标以促进应对气候变化目标的潜在可能性。另外，随着区域性、全球性风险的影响逐渐加深，单一国家发展目标的实现越来越依赖全球公共物品的供应，尤其是在应对气候变化领域，而更具全球包容性的一体化公共物品供给

---

① 根据 WRI. The Future of the Funds：Exploring the Architecture of Multilateral Climate Finance [R/OL]. https://www.wri.org/sites/default/files/The_Future_of_the_Funds_0.pdf 修改。

② Delina L. Multilateral development banking in a fragmented climate system: shifting priorities in energy finance at the Asian Development Bank [J]. International environmental agreements: politics, law and economics，2017，17 (1): 73-88.

机制仍然缺位①。

在高度依赖 ODA 的气候资金体系中，来自发达国家的公共资金规模相对固定，金融危机后，国际金融机构为了实现既定的绩效目标，更倾向于见效快、规模大、投资回报收益相对较高的减缓类项目。在过去的 15 年里，全球气候资金机构在气候领域投入了 117 亿美元资金，共支持了 697 个项目，撬动私人和其他渠道的资金 700 亿美元。但其中大部分基金仍然倾向于投资大型项目，而非最脆弱地区或最脆弱人群急需的小型项目。在 GCF 现有的项目中，有 90% 以上的支出是通过联合国发展署（United Nations Development Program，UNDP）、世界银行以及泛美开发银行等机构，而非通过最迫切需要支持的国家或区域型的机构来发放，CIF 最近发布的 3 年计划表明这一趋势仍在加强。这些机构现有的绩效管理体系中大多缺乏对最脆弱地区、最脆弱人群等全球公共目标投资的机制设计和绩效目标，这也是致使适应、能效、最脆弱地区的小型项目始终很难获得资金保障的重要原因。

（三）资金机制之间的协同度有待提高

经过多年的发展，气候适应资金及工具都有所增长，然而即使在 UNFCCC 框架下形成的基金渠道也存在协同度有待改进的问题。例如，适应基金（Adaptation Fund，AF）和最不发达国家基金（Least Developed Country Fund，LDCF）GCF 主要支持小规模适应项目，项目支持范围存在一定程度的重叠。AF 项目审批周期短，在直接向发展中国家的机构提供资金方面积累了多年经验，但 AF 的融资成本相对较高，资金来源不稳定，资金额度较低，只能支持小规模的适应活动，很难保障项目的可持续性。LDCF 和 GCF 则更适合专注于有协同效应，并长期支持的大型项目，但 GCF 要求机构证明项目有额外融资的能力，对风控要求高，也存在如法律和管理体系不适应、缺乏

---

① 胡鞍钢，张君忆，高宇宁. 对外援助与国家软实力：中国的现状与对策［J］. 武汉大学学报：人文科学版，2017（3）：5-13.

行之有效的适合最不发达地区的商业模式等具体的障碍①②③。

（四）新旧多边发展机构需要寻求协同机制与模式创新

多边开发银行在减缓与适应气候变化领域的角色仍值得期待。全球六家最大的多边开发银行（Multilateral Development Banks，MDBs）联合发布的报告显示，2015年多边开发银行共承诺了250亿美元左右的气候资金，该数据包括了多边开发银行的内部资金和通过多边开发银行渠道流入发展中国家的外部资金④，相较2014年的257亿美元，有小幅下降，但气候相关项目的投资总量高于化石燃料⑤⑥。此外，2015年MDBs承诺到2020年将在气候相关项目中每年增资近400亿美元，即使如此，该数字也只占MDBs总投资额不到20%⑦。而事实上，全球公共资本流动与实现可持续发展目标资金需求之间的差距永远不会通过公共资金得到填补，这也正在推动多边发展机构绩效考核的方式发生转变，以往的多边机构主要以提供和支付的资金量来衡量其绩效，而近期则越来越多的通过公共资金撬动私营部门和受捐国内部资源的程度来考察其绩效。现有的多边发展银行是一项全球性的资产，可以也应该通过相互协作，以及进一步与新兴机构合作，在面对全球挑战方面发挥更大的作用。

近年来，新的发展援助渠道不断涌现，慈善基金和非政府组织的援助规

---

① 刘靖. 全球援助治理困境下重塑国际发展合作的新范式［J］. 国际关系研究，2017，4：3.

② Strange A M, Dreher A, Fuchs A, et al. Tracking underreported financial flows: China's development finance and the aid – conflict nexus revisited［J］. Journal of conflict resolution，2017，61（5）：935 – 963.

③ Delina L. Multilateral development banking in a fragmented climate system: shifting priorities in energy finance at the Asian Development Bank［J］. International environmental agreements: politics, law and economics，2017，17（1）：73 – 88.

④ 统计的多边开发银行数据主要为世行集团（WBG），非洲开发银行（AFDB），亚洲开发银行（ASDB），美洲开发银行（IADB），欧洲投资银行（EIB）和欧洲复兴开发银行（EBRD）六家最大的多边开发银行，Joint report on MDB，2015.

⑤ UNFCCC. Compilation and synthesis of the biennial submissions from developed country Parties on their strategies and approaches for scaling up climate finance from 2014 to 2020［EB/OL］. http://unfccc.int/resource/docs/2016/sbi/eng/inf10.pdf.

⑥ E3G. Greening financial flows, what progress has been made in the development banks? ［R/OL］. https://www.e3g.org/library/greening – financial – flows – what – progress – has – been – made – development.

⑦ WRI. The Future of the Funds: Exploring the Architecture of Multilateral Climate Finance［R/OL］. https://www.wri.org/sites/default/files/The_Future_of_the_Funds_0.pdf.

模有所增长，气候领域私有商业化资金流已大幅超越政府资金。新兴市场国家可为不发达的邻国提供官方金融和科技支持，新兴多边组织以及区域发展银行在援助资金规模上已经超过以前以传统多边发展银行为主导的多边援助框架。例如，中国目前对非洲援助金额就已超过世界银行，从已有的统计数据来看，中国投资设立和中外合资的基金共14只，总规模近1400亿美元。这些基金主要投资于发展中国家，包括非洲地区、拉丁美洲地区，"一带一路"沿线国家与东盟国家。除2只绿色主题基金——气候变化南南合作基金与中美绿色基金以外，其他非绿色主题基金也开始践行绿色投资理念。且新兴市场国家在针对周边的气候脆弱国家与地区也积累了非常多的当地案例与经验，进行气候援助与投资方面逐渐显现优势。

与此同时，传统多边机构极其制度与机制设计亟待创新与改进。首先，是目前的气候投资过于保守，总体上传统多边机构的贷款与资本比率基本维持在4:1以下，极大地限制了其资金供给的能力，一般商业银行贷款与资本比率维持在10:1左右，事实上，7:1在国际上也被认为是较为合宜的比例。根据ODI的研究，倘若多边开发银行将这一比例提高至5:1，将可额外释放2000亿美元资金，若提高至7.5:1则可以释放3800亿美金。可为全球已经极度稀缺的气候资金体系注入新鲜血液。其次，多边基金审批项目的流程规则的复杂性也是目前阻碍其在气候融资领域发挥更大作用主要障碍之一。不同基金审批项目的规则不同，申请资金的国家须投入大量时间精力学习各个基金的制度规则，极大地降低了效率。因此，多边气候基金采取统一的申请要求及保障性措施具有极大的必要。但多边金融机构有强大的研究能力和行业政策视野，既可以提供成功与失败的全球案例，又能够促进多层次、多相关方开展对话，厘清公共、私人资金以及多层次机构的优势与角色功能，这对发展中国家来说弥足珍贵[1]。

《巴黎协定》之后，新旧多边机构面临着竞争与合作的全新格局，一方面，大部分发展中国家都在经历高速发展的过程，依然需要通过多边气候融资机制得到国际资本市场融资渠道的支持。另一方面，发达国家和新兴市场国家都需要通过多边机构加强全球、区域性风险管理的能力，以防止全球公

---

[1] Becault E, Marx A. The global governance system for climate finance: towards greater institutional integrity? [EB/OL]. https://ghum.kuleuven.be/ggs/publications/working_ papers/2016/173becault.

共物品供给不足导致危害已经积累形成的国家发展成果。例如，亚投行、南南基金、丝路基金等区域机构则需要专注于制定公共产品和服务的区域协议，例如支持东盟经济共同体、区域基础设施走廊、亚洲碳市场计划等。新兴市场周边与援助国家/地区的气候投资，需要更多地结合新兴市场国家援助、对外投资等长期积累的本地经验①。因此，如何推进新旧多边机构在新的格局中发挥优势，协同创新也是气候融资体系发展的重要议题。

案例 1

加勒比巨灾风险保险基金

2016 年 9 月下旬，5 级飓风马修登陆东加勒比地区造成强降雨对圣卢西亚、牙买加、海地等国家造成了大面积影响。其中圣卢西亚大部分地区都发生了洪涝灾害，部分地区发生山体滑坡，多达 85% 的农场遭受损失。圣卢西亚是加勒比巨灾风险保险基金（CCRIF）的参与国之一，据测算，该地区降雨损失指数超过 CCRIF 的飓风及洪涝保险政策的启动水平，在飓风发生的 14 天内，CCRIF 即向圣卢西亚赔付 3781788 美元。此外，31 位圣卢西亚居民因参保加勒比气候风险适应和保险计划下的生计保障产品，在此次飓风灾害后得到总额为 102000 美元的保险赔款。生计保障计划是 CCRIF 专为低收入人群设计提供的应对极端天气风险的微型保险产品。

CCRIF 于 2007 年在加勒比地区注册成立，是一个为加勒比地区政府（2015 年后也开始为中美地区提供服务）提供极端天气及地震自然灾害风险管理服务的资金机制。CCRIF 下设三个项目：加勒比综合主权风险管理项目、加勒比气候风险适应和保险计划、气候适应经济项目。CCRIF 旨在通过向受飓风、洪涝或地震影响的加勒比地区政府提供迅速及时的短期流动性资金来降低受灾地区政府的财务损失，是世界上迄今为止唯一使用参数保险的区域性基金，能使加勒比各国政府以最低的价格购买飓风和地震灾害保险。CCRIF 的设立代表了政府处理自然灾害风险方式的转变。

---

① Dorsch M J, Flachsland C. A polycentric approach to global climate governance [J]. Global environmental politics, 2017, 17 (2): 45 – 64.

CCRIF 最初的资金由日本政府资助，后续资金来源于加拿大、欧盟、英国、法国、爱尔兰、百慕大政府、世界银行和加勒比开发银行捐助的信托基金，以及参与国政府支付的会员费。为了保障 CCRIF 的偿付能力，CCRIF 将各个国家的风险放入一个风险池，并进行了大量的建模工作，确保基金可以为一年内发生概率不超过万分之一的事件提供赔付。概率为万分之一的事件发生基本上意味着在加勒比地区的一些大型经济体会同时发生巨大的灾难，例如在一年内牙买加、巴巴多斯、特立尼达、开曼和巴哈马同时受到自然灾害的冲击。尽管这类事件发生的可能性很小，但 CCRIF 在其财务中也将这类可能性囊括在内，并可以为此类小概率集体灾害性事件进行赔付。

案例 2

非洲风险防范能力机制

据估算，非洲撒哈拉沙漠以南地区由于旱灾发生而可能导致的直接经济损失高达 30 亿美元，① 这对于救助国家和当地国家政府的财政都是巨大负担。

马拉维在 2015 年至 2016 年期间遭遇了严重干旱，全国因为粮食短缺问题而陷入灾难状态，但马拉维曾向非洲风险防范能力机制（ARC）购买了应对极端天气的保险，最终 ARC 基于干旱造成的影响和保险合同覆盖的范围，向马拉维支付了 810 万美元的赔款，弥补了部分财务损失。在该案例中，虽然保险资金由于技术原因比原定的到位时间有所延迟，但整体而言大幅减轻了政府的财政负担。

ARC 是非洲联盟为了帮助成员国提高规划、准备和应对极端天气事件与自然灾害的能力于 2012 年特别成立的资金机制。目前 ARC 下设三个产品：极端天气防范机制（Extreme Climate Facility，XCF）、传染性疾病暴发防范机制（Outbreak & Epidemic，OE）和复制性覆盖机制（Replica Coverage，RC）。

---

① ARC 官网 http://www.africanriskcapacity.org/2016/10/29/vision-and-mission/.

ARC通过使用联合国世界粮食计划署开发的一款先进的卫星天气监视软件——非洲风险评估（Africa Risk View），来量化评估遭受恶劣天气影响的非洲国家情况并根据相应的机制进行保险赔付。该资金池的初始资本来自参与国的保费以及一次性合作伙伴的捐助。各国可根据自身需要的干旱风险保障水平从ARC选择相应等级的费率。ARC对这笔风险资金进行了再保险，同时也使用保险资金进行投资活动。当灾害发生时，ARC根据预先制定的透明的付款规则向成员国进行赔付。由于通常而言灾害不会在同一年于非洲所有地区同时爆发，因而建立像ARC这样的灾害风险池不仅可以大幅降低非洲各国应急的成本，同时对外援的依赖也可大幅减少。

二、中国气候融资进展

近年来,随着气候变化政策持续推进、政府与社会资本合作(Public-Private Partnership, PPP)模式在绿色领域广泛应用以及绿色金融体系建设逐步落地,中国的气候资金渠道不断丰富。同时,通过与其他发展中国家合作设立绿色基金,中国的绿色投资实践开始走向国际,不仅向国际进行气候资金输出,也进一步丰富了国内的气候资金来源。

(一)气候变化政策体系逐步完善

继2007年国务院发布《中国应对气候变化国家方案》以来,中国政府相继采取了一系列措施积极应对气候变化,逐步建立起了覆盖减缓与适应两方面的气候变化政策体系,为国内的气候融资供给提供了有力的政策支持。

1. 碳排放强度控制

"十二五"规划首次将单位国内生产总值碳排放强度作为约束性指标纳入发展规划中,标志着低碳转型已成为中国社会经济发展的重要目标之一。国务院随后发布的《"十二五"控制温室气体排放工作方案》将单位GDP减排任务下分至31个省级政府,并出台了系列文件推动温室气体统计常态化与规范化,以及时考核评估各省级政府的单位GDP碳减排任务完成情况。在国家政策指引下,各省级政府进一步将碳排放强度控制责任细分至各市级政府。"十三五"时期,单位GDP碳强度控制继续被纳入《"十三五"规划纲要》的目标体系,自此中国已逐渐形成以行政区域划分的、多层级的碳排放强度控制制度。

2. 碳交易市场

2011年11月,国家发改委批准北京、上海、天津、重庆、广东、湖北、深圳七省市开展全国碳交易市场试点。自2013年起,各试点碳市场相继启动,现已初步形成了符合地区实际的碳交易制度,碳交易市场初具规模。截

二、中国气候融资进展

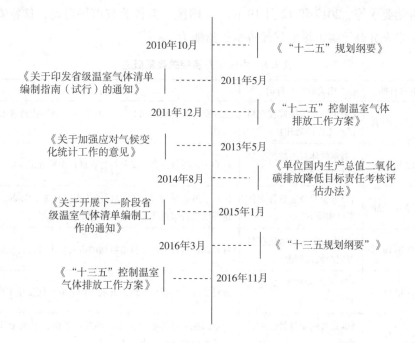

图 2-1 碳强度控制政策概览

至 2016 年年底，中国七个碳交易试点累计成交量 1.16 亿吨，累计成交金额近 25 亿元人民币，2016 年成交金额为 10 亿元，较 2015 年增长 22.1%[①]。根据各交易所的公开信息，2016 年中国抵消碳信用 CCER 累计交易量约 5300 万吨二氧化碳当量，已备案 CCER 项目共计 861 个，和 2015 年相比，2016 年新增备案项目 409 个。截至 2017 年底七个试点累计成交量 1.82 亿吨，累计成交额超过 36 亿元[②]，2017 年成交金额约为 11 亿元，较 2016 年增长了 10%，显示出国内碳市场的快速发展态势。

2014 年，国家发改委开始启动全国碳市场的制度设计研究。同年 12 月发布的《碳排放权交易管理暂行方法》对全国碳市场的发展方向、组织架构体系等提出了诸多规范性要求，2016 年发布的《关于切实做好全国碳排放权交易市场启动重点工作的通知》进一步明确了全国碳市场的覆盖行业以及具

---

① 北京环境交易所、北京绿色金融协会．（2017）．北京碳市场年度报告 2016. http：// www.cbeex.com.cn/article/xxfw/xz/bjtscndhq/201701/20170100059897.shtml.
② 北京环境交易所、北京绿色金融协会．（2018）．北京碳市场年度报告 2017. http：// files.cbex.com.cn/cbeex/201802/20180211162427630.pdf.

体启动要求等。2017年12月19日,全国统一碳排放权市场启动,这将有助于未来充分释放碳市场的气候资金供给能力。

表2.1 中国碳交易相关政策概览

发布日期	相关政策与行动	相关内容
2011年10月	《关于开展碳排放权交易试点工作的通知》	批准京津沪渝粤鄂深七省市2013年开展碳排放权交易试点
2012年6月	《温室气体自愿减排交易管理暂行办法》	对CCER项目开发、交易与管理进行系统规范
2012年7月	《万家企业节能目标责任考核实施方案》	拉开了节能量交易市场建设的序幕,山东、江苏、福建等已先行试点建立全省的节能量交易市场
2012年10月	《温室气体自愿减排项目审定与核证指南》	对CCER项目审定与核准机构的备案要求等进行规定
2013年11月	《国家适应气候变化战略》	应对全球气候变化,统筹开展全国适应气候变化工作
2014年12月	《碳排放权交易管理暂行方法》	对全国统一碳排放权交易市场发展方向、组织架构设计等提出规范性要求
2015年9月	《生态文明体制改革总体方案》	深化碳排放权交易试点建设,逐步建立全国碳排放权交易市场
2015年9月	《中美元首气候变化联合声明》	2017年启动全国碳排放交易体系,将覆盖钢铁、电力、化工、建材、造纸和有色金属等重点工业行业
2016年1月	《关于切实做好全国碳排放权交易市场启动重点工作的通知》	明确参与全国碳市场的8个行业,要求对纳入企业历史碳排放进行核查、提出企业碳排放补充数据核算报告等
2016年1月	《关于试行可再生能源绿色电力证书核发及自愿认购交易制度的通知》	将从2018年适时启动
2016年8月	《新能源汽车碳配额管理办法》征求意见稿	拟运行时间待定
2016年11月	《"十三五"控制温室气体排放工作方案》	对"十三五"时期应对气候变化、推进低碳发展工作作出全面部署
2016年12月	关于发布《绿色发展指标体系》《生态文明建设考核目标体系》的通知	碳减排作为生态文明建设评价考核的依据

续表

发布日期	相关政策与行动	相关内容
2016年12月	《"十三五"节能环保产业发展规划》	发展节能环保产业,加强污染的防治工作
2017年1月	《中华人民共和国气候变化第一次两年更新报告》	①全面更新了国家温室气体清单 ②系统总结和分析了我国"十二五"期间的减缓行动及其效果 ③对资金、技术和能力建设需求及获得的资助进行了更新 ④首次报告了国内测量、报告和核查(MRV)体系
2017年8月	《用能权有偿使用和交易制度试点方案》	明确从2017年起在浙江、福建、河南、四川省开展试点,2020年视情况逐步推广。目前,四个试点省份均处于制度设计阶段
2017年12月	《全国碳排放权交易市场建设方案(发电行业)》	明确用一年时间完成基础建设;一年时间完成模拟运行,在发电行业交易主体之间开展配额现货交易;在发电行业碳市场稳定运行的基础之上,扩大市场覆盖范围与交易品种和交易方式

3. 多层级低碳试点

"十二五"期间,中国积极开展各类低碳试点,已初步形成了覆盖省区、城市、工业园区、城镇、社区等多层级低碳试点格局。

从2010年起,国家发改委开始大力推进低碳省区与低碳城市试点工作。至2017年,已先后开展了三批低碳省市试点,覆盖29个省区81个城市,试点工作在全国范围内全面铺开。自2012年公布第二批低碳省区试点名单以来,中国开始要求各试点确立相应的排放峰值目标以形成倒逼机制,截至2017年10月,已有34个省市提出了实现碳排放峰值的年份目标,其中北京、上海、广州、杭州等12个城市明确提出了2020年前达到碳排放峰值。据相关评估显示,"十二五"期间低碳省市试点地区的单位GDP碳排放下降幅度明显优于全国平均水平,低碳试点已初显成效[①]。

为探索工业的低碳发展新模式,2013年9月,工信部与国家发改委联合发布了《关于组织开展国家低碳工业园区试点工作的通知》,自此中国的低碳工业园区试点工作正式启动。目前,已有51家低碳工业园区获批,在试点期间其

---

① 清华-布鲁金斯公共政策研究中心.中国低碳发展报告2017[R].

单位工业增加值的能源消耗和碳排放均显著下降，碳生产力得到有效提升[①]。

低碳城镇方面，2015年8月，国家发改委印发《关于加快推进国家低碳城（镇）试点工作的通知》，确定了广东深圳国际低碳城、广东珠海横琴新区等8个城镇作为首批国家低碳城（镇）试点，旨在从规划、建设、运营、管理等全周期探索新型城镇化过程中的低碳发展模式。

2014年3月国家发改委发布的《关于开展低碳社区试点工作的通知》拉开了低碳社区试点的序幕。2015年2月，国家发改委印发《低碳社区试点建设指南》，进一步指导了城市新建社区、城市既有社区、农村社区等不同类型社区的低碳试点工作，推进了以社区为单位控制碳排放的路径探索。目前，全国范围内已开展了800个左右的低碳社区试点[②]。

4. 适应城市试点

除了推动低碳减排试点外，中国政府对于适应气候变化的重视程度不断加大，自1994年颁布的《中国二十一世纪议程》中首次提出适应气候变化的概念后，2010年在《十二五规划》中，明确提出要求："在生产力布局、基础设施、重大项目规划设计和建设中，充分考虑气候变化因素。提高农业、林业、水资源等重点领域和沿海、生态脆弱地区适应气候变化水平"。2013国家发展改革委、财政部、住房城乡建设部、交通运输部、水利部、农业部、林业局、气象局、海洋局联合制定了《国家适应气候变化战略》。

为积极探索适应气候变化的城市解决方案，2016年1月国家发改委联合住建部共同制定了《城市适应气候变化行动方案》，从城市规划、基础设施、建筑、生态绿化、水、风险管理等角度提出了一系列发展目标与任务，各地就此开展了积极探索。2017年2月，两部门再次联合出台了《关于印发气候适应型城市建设试点工作的通知》，明确了将适应气候变化理念纳入城市规划建设管理全过程的工作目标，确定了内蒙古呼和浩特、辽宁大连、湖北武汉等28个地区作为首批气候适应型城市建设试点，从城市适应理念、气候变化和气候灾害监测预警能力、重点适应行动、政策实验基地和国际合作平台五个方面明确试点城市到2020年的工作任务。自此，我国适应城市试点工作正式启动。

---

① 中国社会科学院城市发展与环境研究所. 国家低碳工业园区建设实践与创新[R].
② 清华-布鲁金斯公共政策研究中心. 中国低碳发展报告2017[R].

二、中国气候融资进展

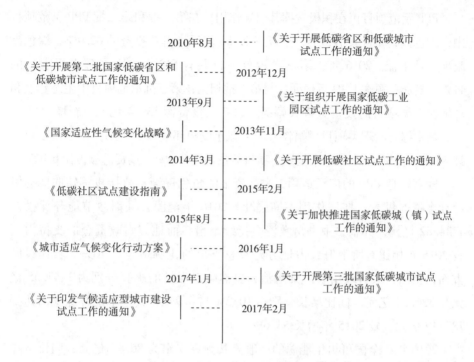

图 2-2 低碳试点的相关政策与行动

（二）PPP 模式在绿色低碳领域的运用进一步深化

近年来，政府与社会资本合作模式（PPP）的政策支持力度不断增强，尤其是针对绿色低碳领域。2016 年 10 月财政部出台的《关于在公共服务领域深入推进政府和社会资本合作工作的通知》指出，在城市轨道交通、清洁能源设施、垃圾处理、污水处理等公共服务领域，由于其市场化程度高、PPP 模式运用成熟，各地新建项目应"强制"应用 PPP 模式。

根据财政部 PPP 项目信息库数据，截至 2017 年 12 月末，管理库项目共计 7137 个，累计投资额 10.8 万亿元，季度环比净增项目 359 个，投资额 6376 亿元；年度同比净增项目 2864 个，投资额 4.0 万亿元，其中，处于执行和移交阶段的项目（已落地项目）2729 个（目前移交阶段项目 0 个），投资额 4.6 万亿元，落地率 38.2%（即已落地项目数与管理库项目数的比值）。污染防治与绿色低碳领域共有项目 3979 个，投资额 4.1 万亿元，分别占管理库的 55.8% 和 38.0%，年度同比净增项目 1507 个，投资额 1.4 万亿元。

PPP 示范项目正在积极发挥其引导作用，在第一批和第二批 PPP 示范项目中，绿色低碳项目共 105 个，投资 5049.56 亿元，占总投资的 62.9%，绿色市政项目是主流。2016 年，第三批示范项目共有 513 个，占全部申报项目的 44%，投资总额逾 1.19 万亿元。与第二批相比，第三批示范项目中生态建设和环境保护类项目的投资金额大幅提升，新增投资额 539.4 亿元，新增项目 33 个，其中综合治理类项目增幅较大。生态建设和环境保护类第三批示范项目共计 46 个，投资总额 810.56 亿元，项目数量占比 8.9%，投资总额占比 6.9%。

随着绿色 PPP 的广泛运用，财政资金的角色逐渐从直接投资转向撬动引导社会资本投入，杠杆作用不断强化。2016 年全国公共财政节能环保投入 4734.82 亿元，其中，中央财政资金持续下滑，而地方财政资金的支持力度较 2015 年相比有所上升，占地方财政总支出的比例增至 2.76%。具体投向方面，全国公共财政用于污染防治与自然生态保护的决算分别为 1447.55 亿元与 326.54 亿元，同比增长分别达 10.2% 与 6.9%；其他节能环保支出达到 787.49 亿元，较 2015 年增长 53.6%。

2016 年，全国可再生能源的一般公共预算支出为 86.12 亿元，同比下降 47.7%，其中主要为地方财政预算支出减少，而中央在该领域的公共预算支出大幅上升，较上年增长 418.1%；此外，全国政府性基金在可再生能源电价附加收入安排的支出达 595.06 亿元，较 2015 年上升 2.7%。

表 2-2　中央财政和地方气候变化支出与其他项目的比较

单位：亿元

年份	中央			地方		
	2014	2015	2016	2014	2015	2016
节能环保	2033.03	400.41	295.49	3470.90	4402.48	4439.33
教育	4101.59	1358.17	1447.72	21788.09	24913.71	26625.06
科学技术	2541.81	2478.39	2686.10	2877.79	3384.18	3877.86
文化体育与传媒	508.47	271.99	247.95	2468.48	2804.65	2915.13
医疗卫生	2931.26	84.51	91.16	10086.56	11868.67	13067.61
总支出	74161.11	80639.66	86804.55	151785.56	175877.77	160351.36
节能环保占比（%）	2.74	0.5	0.34	2.29	2.50	2.76

资料来源：财政部预算司财政数据。http://yss.mof.gov.cn/2016js/.

二、中国气候融资进展

表2-3 PPP示范项目绿色低碳相关行业统计表

领域	第一批示范项目		第二批示范项目		第三批示范项目		合计			
	项目数	投资额（亿元）	项目数	投资额（亿元）	项目数	投资额（亿元）	项目数	比例（%）	投资额（亿元）	比例（%）
1. 交通：	9	1565.96	38	3491.78	40	4467.48	87	11.48	9525.22	47.98
轨道交通	7	1526.51	13	2452.67			20	2.64	3979.18	20.05
高速公路			7	613.94	26	3689.36	33	4.35	4303.30	21.68
非收费公路			8	232.38			8	1.06	232.38	1.17
交通枢纽			4	17.32	2	20.57	6	0.79	37.89	0.19
机场			2	81.98	1	203.17	3	0.40	285.15	1.44
铁路			2	45.7	3	126.41	5	0.66	172.11	0.87
公交			1	15			1	0.13	15.00	0.08
桥梁			1	32.78	4	370.09	5	0.66	402.87	2.03
其他	2	39.45			4	57.87	6	0.79	97.32	0.49
2. 市政工程：	8	77.27	66	810.53	119	1495.35	193	25.46	2383.15	12.01
垃圾处理	1	5.26	22	97.24	31	124.44	54	7.12	226.94	1.14
地下综合管廊	1	13	14	407.77	31	838.68	46	6.07	1259.45	6.34

续表

领域	第一批示范项目 项目数	第一批示范项目 投资额（亿元）	第二批示范项目 项目数	第二批示范项目 投资额（亿元）	第三批示范项目 项目数	第三批示范项目 投资额（亿元）	合计 项目数	合计 比例（%）	合计 投资额（亿元）	合计 比例（%）
公园			1	25	4	43.94	5	0.66	68.94	0.35
供气			2	3.64	2	3.40	4	0.53	7.04	0.04
供热	3	26.4	6	59.52	13	62.24	22	2.90	148.16	0.75
供水	3	32.61	18	190.11	24	152.65	45	5.94	375.37	1.89
海绵城市			1	13.85	5	208.90	6	0.79	222.75	1.12
绿化			1	5.5	4	16.22	5	0.66	21.72	0.11
排水			1	7.9	5	44.88	6	0.79	52.78	0.27
3. 环境保护:	11	111.99	31	742.46	82	932.79	124	16.36	1787.24	9.00
污水处理	9	59.08	15	337.49	40	181.37	64	8.44	577.94	2.91
环境综合治理	2	52.91	15	364.97	38	711.32	55	7.26	1129.20	5.69
湿地保护			1	40	4	40.10	5	0.66	80.10	0.40
合计	28	1755.22	135	5044.77	241	6895.62	404	53.30	13695.61	68.99

资料来源：财政部政府和社会资本合作中心 PPP 项目库。

（三）绿色金融工具撬动气候资金能力不断显现

2016年到2017年，中国的绿色金融体系建设持续全面推进，各项绿色金融工具的市场规模进一步扩大，应对气候变化的资金渠道不断拓宽。

1. 绿色信贷

作为中国绿色金融体系中起步最早的领域，绿色信贷也是目前中国最重要的绿色投融资渠道之一。截至2016年年底，中国21家主要银行金融机构的绿色信贷余额达到7.51万亿元，同比增长7.13%，占各项贷款余额的8.83%。其中，节能环保项目和服务的贷款余额为5.81万亿元，节能环保、新能源、新能源汽车等战略新兴产业贷款余额为1.7万亿元，预计可年节约标准煤1.88亿吨，减排二氧化碳当量4.27亿吨。资金投向方面，绿色交通运输和清洁能源项目是中国绿色信贷最主要的投向领域[1]。

中国绿色信贷产品创新层出不穷，绿色资产的金融杠杆作用不断强化。其中既包括以碳资产、合同能源管理项目未来收益权等绿色权益为抵质押品的贷款产品，也包括直接支持绿色项目的融资产品。此外，绿色信贷资产证券化的发展也将进一步释放资金。

2. 绿色证券

2017年，中国境内外绿色债券发行规模达2483.14亿元，占全球发行规模的32%，与2016年相比有所下降，但依然是全球最大的绿色债券发行国[2]。图2-3为中国绿色债券募集资金用途分布。如图2-3所示，除金融债外，清洁能源是中国近两年绿色债券募集资金投向最多的领域，主要涉及光伏发电、风力发电、水力发电等。此外，中债—中国气候相关债券指数、中财—国证绿色债券指数等衍生品品种也日益丰富。未来，通过开发以绿色债券指数为标的的交易型开放指数基金产品，有利于调动社会更多资金以支持应对气候变化领域的项目。

3. 气候巨灾保险

作为应对气候变化的重要风险分散机制，近年来气候巨灾保险制度建设在中国取得了一系列进展，其在增强气候变化适应能力、防止因灾致贫等方面的功能逐步显现。目前，已有宁波、深圳、厦门等多个城市先后启动了巨

---

[1] 资料来源：中国银监会统计数据。
[2] 中央财经大学绿色金融国际研究院，中国绿色债券市场2017年度总结。

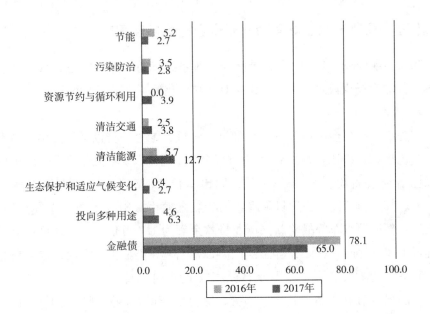

资料来源：中央财经大学绿色金融国际研究院，中国绿色债券市场2017年度总结。

图2-3 中国绿色债券募集资金用途

灾保险机制。以厦门为例，通过财政出资、五家保险公司组成联保体的形式，厦门市全民均获得了针对台风、暴雨、洪水、地震等自然灾害的人身、住房与财产风险保障。

财政巨灾指数保险产品的推出进一步强化了保险在增强地区气候韧性方面的职能。2016年11月，广东省财政巨灾指数保险在10个试点城市实现全面落地，累计提供风险保障23.47亿元；截至2017年8月，已累计支付赔款6527.6万元，在自然灾害救助体系建设中发挥着重要作用。

4. 绿色基金

国家层面，财政部清洁发展机制基金（Clean Development Mechanism Fund, CDMF）作为发展中国家首只应对气候变化政策性基金，自2010年业务全面运行以来，通过社会性基金管理模式，累计投资绿色低碳项目223个，实现减排量4654.62万吨二氧化碳当量①。地方层面，随着国家政策的大力推动，绿色基金已成为推动城市绿色转型的重要资金来源，且新增数量明显

---

① 数据来源：http://www.cdmfund.org/zh/ycsy/index.jhtml.

上升。截至2016年底,中国基金业协会备案的绿色基金共265只,成立于2016年的基金数为121只。在所有备案绿色基金中,绿色产业基金占83%,其中51%为清洁能源产业基金。这为尚未达到绿色信贷、绿色债券等融资工具投资要求的新兴绿色产业提供了重要的资金来源。

(四)中国的绿色投资实践逐步走向国际

近年来,中国的对外绿色投资理念不断深化。继2013年商务部与环境保护部联合发布《对外投资合作环境保护指南》后,2017年4月,环保部出台了《关于推进绿色"一带一路"建设的指导意见》,旨在从基础设施建设、贸易、对外投资等角度贯彻落实生态文明理念,以推进绿色丝绸之路建设,并强调了中方发起的国际多双边合作机构和基金作为绿色资金供给方的重要性。2017年9月,中国金融学会绿色金融专业委员会等七个机构共同发布《中国对外投资环境风险管理倡议》,鼓励和引导中国金融机构加强对外投资中的环境风险管理,遵循责任投资原则。

在政策与行业倡议的推动下,中国积极同各国特别是广大的发展中国家开展双向绿色合作,旨在实现参与国经济转型发展与环境保护的双赢局面。其中,由中方投资设立或中外合资的基金已成为重要的绿色合作平台。

从已有的统计数据来看,中国投资设立和中外合资的基金共14只,总规模近1400亿美元。这些基金最主要投资于发展中国家,包括非洲地区、拉丁美洲地区、一带一路沿线国家以及东盟。除2只绿色主题基金——气候变化南南合作基金与中美绿色基金以外,其他非绿色主题基金也开始践行绿色投资理念,提供了诸多宝贵绿色投资经验。

表2-4 中外合作基金概览

基金名称	成立时间	基金承诺出资规模(十亿美元)	关注的领域	投资地理范围
亚洲地区				
丝路基金	2014	54.5	基础设施建设、能源、产能	"一带一路"沿线国家,以亚洲国家为主
中国—东盟投资合作基金	2009	10	基础设施建设、能源、自然资源	中国、东盟

续表

基金名称	成立时间	基金承诺出资规模（十亿美元）	关注的领域	投资地理范围
中国—东盟海上合作基金	2011	0.5	海运经济、环境保护	中国、东盟
欧亚合作				
中国—中东欧投资合作基金	2012	11.5	基础设施建设、能源、制造业、通信	中东欧
俄中投资基金	2012	1（俄罗斯）+ 1（中国）= 2	基础设施建设、农业、自然资源	70% 投资于俄罗斯，30% 投资于中国
拉美地区				
中国—拉美产能合作投资基金	2014	20	基础设施建设、能源、自然资源、制造业、信息与通信技术	拉美国家
中国—拉美基础设施专项基金	2014	10	基础设施建设	拉美国家
中国—拉美合作基金	2014	5	基础设施建设、能源、自然资源、农业、制造业、信息技术	拉美国家
中国—墨西哥投资基金	2014	2.4	基础设施建设、汽车产业	墨西哥
非洲地区				
中非发展基金	2007	3	采矿、能源、制造业	非洲
非洲开发基金	2014	1（非洲发展银行）+ 1（中国）= 2	基础设施建设	非洲
中非工业能力合作基金有限公司	2015	10	基础设施建设、能源、制造业、农业、采矿	非洲
全球性				
气候变化南南合作基金	2015	3.2	适应与减缓气候变化	发展中国家
南南合作援助基金	2015	2	无特定主题	最不发达国家、小国家、岛国
北美地区				
中美绿色基金（原名：中美建筑节能与绿色发展基金）	2016	3.05（首批募集）	建筑节能、减排、产业结构升级	中国（与地市级政府合作）
基金总数：14 只		总出资规模：138.2 亿美元	最受关注的领域：基础设施建设、能源合作	主要关注的国家：发展中国家

二、中国气候融资进展

1. 区域经济一体化目标下的绿色投资：丝路基金

丝路基金于 2014 年 12 月 29 日由外汇储备、中国投资有限责任公司、国家开发银行和中国进出口行共同出资在北京成立，资金规模为 400 亿美元，2017 年中国宣布新增资金 1000 亿元人民币。作为中长期开发投资基金，丝路基金旨在通过股权、债券、贷款在内的多元化金融工具，推进"一带一路"沿线国家的基础设施、资源开发、产业合作、金融合作等领域项目。截至 2017 年 3 月，丝路基金已签约 15 个项目，承诺投资金额累计约 60 亿美元，涉及项目总投资额超过 800 亿美元。

通过将环境保护等社会责任纳入项目可行性评估和风险管理体系，丝路基金在推动互联互通的同时，始终贯彻着绿色投资理念。现已签约的绿色项目包括中巴经济走廊卡洛特水电站项目、亚马尔液化天然气一体化项目、阿联酋迪拜哈翔清洁燃煤电站项目等。表 2-5 为丝路基金的绿色投融资大事记概览。

表 2-5 丝路基金绿色投融资大事记

2014 年	11 月 8 日	习近平主席在加强互联互通伙伴关系对话会上宣布，中国将出资 400 亿美元成立丝路基金
	12 月 29 日	丝路基金在北京正式注册成立
2015 年	4 月	丝路基金投资中巴经济走廊优先实施项目之一卡洛特水电站
	12 月 14 日	丝路基金与哈萨克斯坦出口投资署签署《关于设立中哈产能合作专项基金的框架协议》
	12 月 17 日	丝路基金与俄罗斯诺瓦泰克公司签署亚马尔液化天然气一体化项目交易协议
2016 年	1 月 19 日	丝路基金与沙特国际电力和水务公司签署了共同投资开发阿联酋及电站的谅解备忘录
	6 月 18 日	丝路基金与塞尔维亚政府签署关于新能源项目合作的谅解备忘录
	10 月 13 日	丝路基金与 IFC 亚洲新兴市场基金管理公司签署了关于 IFC 亚洲新兴市场基金项目的认购协议

续表

2017年	5月14日	中国将加大对"一带一路"建设资金支持,向丝路基金新增资金1000亿元人民币
	6月8日	中国与哈萨克斯坦政府签署关于中哈产能合作基金在哈萨克斯坦进行直接投资个别类型收入免税协议
	11月	与通用电气旗下GE能源金融服务在北京签署成立"能源基础设施联合投资平台合作协议",共同投资包括"一带一路"沿线国家和地区的电力电网、新能源、油气等领域基础设施项目

2. 绿色产能合作：中国—拉美产能合作基金

中国—拉美产能合作基金成立于2015年，由外汇储备、国开行共同出资，资金总规模300亿美元，首期资金规模100亿美元。通过推动与拉美国家在制造业、高新技术、农业、能源矿产、基础设施等领域开展产能合作，该基金旨在实现中国与拉美地区的发展互补。

在目前已运作的项目中，不乏绿色低碳领域产能合作的良好实践。以巴西水电站项目为例，中拉产能合作基金与三峡集团联合发起设立特殊目的实体（Special Purpose Vehicle，SPV），竞标投资巴西朱比亚和伊利亚2座水电站（合计装机容量约5GW）30年特许经营权。中国—拉美产能合作基金投资6亿美元，占股比33%，并在很短的时间内撬动银团贷款等外部融资，使得SPV得以如期中标并签署特许经营协议。目前项目已进入投后阶段，运行平稳，预期收益良好。

3. 针对发展中国家的气候援助：中国气候变化南南合作基金

2015年9月，中国国家主席习近平访美期间，正式宣布中国政府出资成立200亿元人民币建立中国气候变化南南合作基金。该基金旨在支持其他发展中国家应对气候变化、向绿色低碳发展转型，包括增强其使用绿色气候基金资金的能力和气候适应力，严格控制对国内以及国外高污染高排放项目的投资。

2016年，中国开始在发展中国家开展应对气候变化"十百千"合作项目——10个低碳示范区、100个减缓和适应气候变化项目及1000个应对气候变化培训名额。"十三五"期间，中国将加速筹建气候变化南南合作基金，配合推进"一带一路"战略和国际产能合作，积极助力其他发展中国家落实可持续发展议程。

二、中国气候融资进展

表2-6 中国气候融资发展情况

分类		规模	时期（年）	说明	数据来源
中国碳市场	中国碳市场试点	累计配额成交金额近25亿元	2016	截至2016年12月31日，七省市试点碳市场累计成交量1.6亿吨，累计成交额近25亿元。其中，2016年包括福建在内的各省市二级市场接近6400万吨，线上线下共成交碳配额现货约80%；交易额约10.45亿元，较2015年交易总量增长约2015年增长22.1%	《北京碳市场年度报告2016》
	中国抵消信用-CCER	5300万吨二氧化碳当量	2016	截至2016年12月31日，中国自愿减排交易信息平台累计公示CCER审定项目2742个，861个项目表备案，254个项目已产生5300万吨碳减排量	中国自愿减排交易信息平台
慈善资金	中国绿化基金会所获赠款	5445万元	2016	2016年度公益事业支出4243万元	《中国绿化基金会2016年度审计报告》
	中国所接收国内外款物捐赠中流向生态环境领域的数额	76.6亿元	2015	2015年全年生态环境领域共接受捐赠76.6亿元，占总捐赠额6.91%，较2014年增长3.41%	《2015年度中国慈善捐助报告》

25

续表

分类			规模	时期（年）	说明	数据来源
传统金融市场	传统国际金融市场	中国对可再生能源的投资	783亿美元	2016	与2015年相比，总投资额下降了32%，是从2013年起以来的最低水平，但中国仍然是世界上第一大投资国	《2017年可再生能源全球现状报告》
		中国主要银行业金融机构绿色信贷余额	7.51万亿元	2016年末	截至2016年末，21家主要银行业金融机构的绿色信贷余额升至7.51万亿元，占各项贷款余额的8.83%，预计可年节约标准煤1.88亿吨，减排二氧化碳当量4.27亿吨，减排化学需氧量271.46万吨，氨氮35.89万吨，二氧化硫488.27万吨，氮氧化物282.69万吨，节水6.02亿吨	《2016年度中国银行业社会责任报告》
	国内金融市场	中国银行业节能环保贷款余额	5.81万亿元	2016		
		中国银行业战略性新兴产业贷款余额	1.70万亿元	2016		
		国内财政资金中节能环保投入	4743.82亿元	2016	其中：能源节约利用622.65亿元，可再生能源86.12亿元	Wind资讯
		绿色债券规模总量	2380亿元	2016		《中国绿色债券市场现状报告2016》
		非贴标绿色债券发行规模	5520亿元	2016		

二、中国气候融资进展

续表

分类			规模	时期（年）	说明	数据来源
传统金融市场	国内清洁技术领域	中国清洁技术行业获得VC/PE投资规模	53.28亿美元	2016	与2015年相比融资案例数量下降43.88%，融资规模上升91.40%。	投中集团旗下金融数据产品CV-Source
		中国清洁技术行业并购市场完成交易规模	披露交易规模124.77亿美元	2016	完成交易案例156起，披露交易规模190.93亿美元。整体来看，清洁技术并购市场宣布交易案例数量及完成交易案例规模均呈下降趋势，完成交易案例规模有大幅提升	
	国际市场和国内市场	中国清洁技术企业境内IPO融资	3.26亿美元	2016	2016年清洁技术行业仅4家企业实现IPO，与2015年相比较下降55.56%，IPO融资规模3.26亿美元，较2015年骤降87.45%。4起IPO企业中，深交所上市1家，港交所上市3家	
企业直接投资	入库项目	生态建设和环境保护林业	6534亿元 80亿元	2016		财政部PPP项目库
PPP项目	对外发布项目	绿色低碳项目	5.5万亿元	2016	在PPP综合信息平台对外发布的PPP项目中，绿色低碳项目共6612个，总投资5.5万亿元，年度同比净增项目4696个，净增投资额3.53万亿元	
国际多/双边合作机构		中外合作基金	138.2亿美元	2016		公开资料，课题组整理

三、其他发展中国家气候融资进程与合作

新兴市场的崛起给发展中国家合作应对气候变化提供了动能与基础条件,气候融资格局由主要依靠发达国家的援助体系转向发展中国家南南合作等多种援助渠道并存的融资格局。发展中国家开展应对气候变化工作与南南合作的主要原因包括:(1)新兴经济体综合国力与国际地位的发展提升促使其寻求可获得更多话语权与自主权的资金机制;(2)发展中国家对环境变化异常敏感与脆弱;(3)实现部分地缘政治目的;(4)缓解国际气候谈判进程中受到的减排压力;(5)推动先进技术的转移和产品的输出等。

(一)其他发展中国家气候融资取得的进展

1. 气候变化政策体系已经初步建立,气候融资政策体系有待完善

明确的气候目标有助于设定长期轨迹,并向企业和社会发出政治意图信号[1]。清晰的战略性政策信号与框架则可进一步增强投资者信心,起到引导与撬动私人领域资金的作用。发展中国家已经开始初步建立起应对气候变化所需的政策和体制框架。格兰瑟姆气候变化与环境研究所(Grantham Research Institute On Climate Change and the Environment, GRICCE)和全球立法组织(Global Legislators Organization)在2015年全球气候立法的研究[2]中指出,纳入其研究的21个非洲国家中,16个已经制定了国家应对气候变化的政策框架,且除利比亚外,所有非洲国家都制定了解决能源供应问题的气候

---

[1] Michal Nachmany, etc. The 2015 Global Climate Legislation Study [R/OL]. http://www.lse.ac.uk/GranthamInstitute/wp-content/uploads/2015/05/Global_climate_legislation_study_20151.pdf.

[2] 格兰瑟姆气候变化与环境研究所和全球立法组织自2010年起对绝大多数全球温室气体排放国家的气候立法进行研究,是目前覆盖国家数目最广泛且最全面的气候立法研究组织,迄今已经发布7本研究报告。

相关政策[①]，亚太地区和拉丁美洲加勒比地区的大部分国家也已经制定了国家层面的应对气候变化行动框架。可以说几乎所有国家都有某种形式的气候变化立法，但大部分发展中国家的气候变化政策尚未化解为具体行动政策，且缺乏系统的联动性强的气候融资政策体系。

2. 区域资金机制和创新金融模式发展迅速，撬动私人领域资金能力逐步显露

国际社会高度关注建立创新的气候资金平台，通过该类平台创新金融模式与资金工具，在发展中国家建立区域资金机制来撬动私人资本。比较有代表性的包括气候与发展知识网络（Climate and Development Knowledge Network，CDKN），气候融资创新机制（Climate Finance Innovation Facility，CFIF）和全球创新实验室（Global Innovation Lab，GIL）。

CDKN 是一个南北联盟，由英国国际发展署和荷兰外交署捐助成立，并由普华永道牵头的组织机构联盟进行管理。CDKN 提供了 81000 英镑在亚马逊西南部设立森林金融实验室，旨在吸引公共和私营部门的资金用于减少毁林所致排放量，并为向可持续生态系统管理和低碳经济转型提供资金。

CFIF 由联合国环境规划署和法兰克福金融与管理学院共同创立，以支持发展中国家的金融机构参与可再生能源和能源效率领域的投资，该机制向发展中国家金融机构提供技术援助和资金，发展以气候为重点的金融产品以动员更大规模的资金流向减缓和适应领域。目前 CFIF 已经支持了 15 个项目，从中获益的国家包括巴基斯坦、印度、菲律宾、尼泊尔、柬埔寨、蒙古国、新加坡、越南和汤加。

表 3-1　CFIF 支持的 15 个项目

项目	项目地点	重点支持领域
通过建立微型绿色能源公司在巴基斯坦推广可再生能源技术（可再生能源技术）	巴基斯坦	可再生能源，能源效率
绿色住宅和太阳能离网设备的信贷融资	印度	可再生能源，能源效率

---

① AfDB. African INDCs: Investment needs and emissions reductions in the energy sector [EB/OL]. http://www.vivideconomics.com/wp-content/uploads/2016/11/AfDB-Report-2016.pdf.

续表

项目	项目地点	重点支持领域
通过增加获得可再生能源/高效能源来扶持穷人	菲律宾	可再生能源
台州市商业银行能源小企业效益贷款项目	中国	可再生能源，能源效率
为尼泊尔农村贫困社区的家庭太阳能系统提供信贷融资	尼泊尔	可再生能源
农村贫困地区可再生能源电器融资可行性研究	柬埔寨	可再生能源
发展碳融资机制	蒙古国	小额信贷，能源效率
引入沼气	尼泊尔	可再生能源
减缓和适应气候变化	新加坡	能源效率
通过ESCO融资和风险共担计划扩大在中国广东省的能效合同	中国	能源效率
在菲律宾的能源包容倡议	菲律宾	可再生能源，能源效率
清洁能源贷款的设计和试点测试	柬埔寨	可再生能源，能源效率
支持农户的房屋配备太阳能灶和红外炉	越南	可再生能源，能源效率
太平洋能源贷款能力建设	汤加	可再生能源
通过清洁能源产品缓解气候变化	印度	可再生能源，能源效率

　　GIL 由英国、美国、丹麦、法国、日本、荷兰、挪威和德国政府与一些主要发展性金融机构及私营部门的部分参与者于 2014 年合作开发，主要开展资金工具和融资模式的创新示范，成立至今已吸引了 1.7 亿美元的资金。GIL 以 PPP 模式，通过创新金融工具为项目面临的融资挑战提供具体的解决途径。2015～2016 年，GIL 完成了水资源融资机制（Water Financing Facility, WFF）的设计①，其运作机理如下：全球 WFF 将成立有限责任公司，初始资本来自捐赠者和影响力投资者提供的 1.12 亿美元，并将提供股权资本与第一损失担保金来支持符合相关条件的 8 个发展中国家成立国家层面的 WFF。国家 WFF 将为本国的公共或私人水资源设施提供长期低息贷款，并通过资产证券化的方式，将这些贷款打包成资产支持证券（Asset - Backed Security, ABS），向本国机构投资者发行以当地货币计价的、投资级别的债券，水资源设施产生的现金流和剩余权益则仅限于偿还债券的本息。该机制的灵活性在于，除了来自全球 WFF 提供的第一损失担保金外，其他多边机构或当地政府

---

① 详见：https://www.climatefinancelab.org/project/water - finance/.

也可以为国家 WFF 提供额外担保或盈利承诺,以进一步降低投资者风险。当第一个国家 WFF 得以成功运营并到达收支平衡后,全球 WFF 的资金即可收回并用于投资其他国家 WFF 的设立。预计通过上述方式,公共资本与私人资本的杠杆比可达 1:1.4,预计到 2030 年,每年可撬动的私人资金达到 12.3 亿美元。WFF 的具体运作机理如图 3-1 所示。

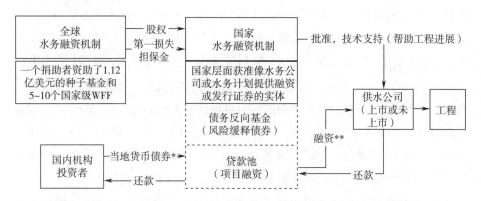

* 注解 1:肯尼亚实行计划的目的是五年时间中用本地货币发行金额为 2.5 亿美元,长达 15~20 年的债券。

** 注解 2:200 万~1000 万美元/100% 项目成本,债券利率传递加上息差。

资料来源:https://www.climatefinancelab.org/project/water-finance/.

图 3-1 WFF 的运作机理

案例 3

肯尼亚 WFF 试点

肯尼亚 WFF 试点已于 2017 年初建立,荷兰外交部称将提供 1000 万欧元的捐赠款,该试点将定期发行水资源和公共卫生项目的债券。此外,目前已筛选了 7 个比较符合条件的发展中国家,可在未来一到两年内复制试点项目。肯尼亚的 WFF 预计将在 2017 年底发行第一只公司债。该试点初步计划通过前八年的债券发行来撬动价值 2.5 亿美元的本地私人资金。通过相关模型量化计算,全球 WFF 的第一损失担保金以及政府和其他多边机构的担保承诺将可覆盖债券 42% 的风险,总共动员的公共资金为 1.2 亿美元,公共与私人资本金的撬动率将达到 1:1.3。

3. 风险管理工具发展迅速，为发展中国家气候适应提供的资金不断增多

目前除了应急信贷以外，很多区域和国家联合发展气候主权保险，用于提高脆弱国家灾后重建能力并减轻政府的财政压力。主权保险是专为政府设计的一类险种，由政府购买，根据自然灾害的严重程度进行赔付，可通过私下安排或参与区域风险共担机制（如CCRIF和ARC）完成。该类保险的优势在于设置严格的参数，一旦参数被激发，即可提供快速赔付，另一个好处是成本确定，可准确预估。CCRIF是针对加勒比地区发展中国家的巨灾保险制度，在这些国家遭遇飓风和地震这类自然灾害损失时，通过巨灾风险保险基金的方式予以赔偿。CCRIF的保费水平大约是商业保险市场的50%；如果某年的指数工具表明会出现严重的某类自然灾害，该项目就会根据事前约定对投保政府进行赔付。在准备金方面，除了多方捐赠和自身的准备金积累外，还通过再保险的方式，与国际再保险市场开展合作，有效分散风险。ARC是非洲联盟为了帮助成员国提高应对诸如干旱等自然灾害的能力而成立的风险防控机制，其运作机制与CCRIF类似。

4. 区域气候基金开始探索地区自主用款权，使资金流入当地最需要支持的项目

资金使用与社区的优先需求相融合容易导致投资过于关注短期目标，存在与国家的长期气候政策目标相冲突的风险，如何在两者间寻求平衡是基金运营者与相关利益方面临的一大挑战。目前发展中国家设立的资金机制已经开始探索将本地需求纳入资金的使用规划，并保持与国家长远战略目标的一致性。埃塞俄比亚的气候韧性绿色经济机制（Climate Resilient Green Economy，CRGE）和肯尼亚的县级气候变化基金（County Climate Change Funds，CCCFs）都在该领域做了有益尝试。瓦吉尔和马瓜尼的CCCF是肯尼亚国家干旱管理局的试点项目，在适应联合会的技术支持下，由英国国际发展部为肯尼亚政府提供650万英镑赠款支持此类项目。该试点旨在促进气候变化在县级发展规划中的主流化，加强县级政府气候适应资金的用款权，并以社区需求为优先考虑。各县将基金经费的20%用于能力建设来支持社区主导的水源治理。其中，在瓦吉尔基金已经投资了十二个区级项目和两个县级项目，总额达480756美元，用于提高用水者可持续利用水源的能力。①

---

① ODI. Decentralising Climate Finance – Insights From Kenya and Ethiopia [EB/OL]．https：//www.odi.org/sites/odi.org.uk/files/resource-documents/11804.pdf.

(二) 发展中国家积极开展气候援助与合作

近些年来，发展中国家不仅自身积极应对气候变化，也开始提供对外气候援助。中国作为最大的发展中国家，在过去十几年对外援助金额飞速增长，对非洲援助金额已超过世界银行，相关文献的研究数据显示，2001年到2013年，中国对外援助额从7.43亿美元上升至74.62亿美元，年均增长率高达21.20%，与同期国际比较，中国国际援助增长率相当于世界平均增长率（10.22%）的2倍、美国（9.30%）的2.28倍、日本（5.35%）的3.96倍、英国（7.17%）的2.96倍、德国（12.32%）的1.72倍[①]，气候援助作为中国对外援助的分支，无疑也在近几年大幅提升。

其他发展中国家也通过合作创立多边机构或对话平台的方式加强气候融资合作。气候政策倡议（Climate Policy Initiative, CPI）追踪的数据显示，2015年到2016年，有大约30亿美元/年的气候资金从发展中国家流向发达国家，另外每年有80亿美元资金在发展中国家之间流动[②]。发展中国家创立的多边机构中比较有名的包括：新开发银行（原金砖国家开发银行）、非洲开发银行、泛美开发银行、亚洲基础设施投资银行。这类多边机构主要通过补贴、贷款、债务、股票、信用额度等直接援助具体项目或当地实施机构，此外，也通过支持受援国的国家政策间接实现气候援助的目标，比如为某国制定气候变化相关的政策提供技术援助或推动环境问题成为国家整体战略规划的主流因素。而发展中国家对国际多边机构的捐款则主要投向全球环境基金和绿色气候基金，虽然发展中国家的捐资额在这两个基金中整体比例不足1%[③]，但极具有象征意义。

(三) 发展中国家的需求及中国对策

1. 发展中国家面临资金缺乏和能力建设的双重挑战

尽管发展中国家在应对气候变化方面取得了较大突破，但仍面临严峻挑

---

① 胡鞍钢，张君忆，高宇宁. 对外援助与国家软实力_中国现状与对策 [J]. 2017, 70 (3). 武汉：武汉大学学报, 2017.
② CPI. The Global Landscape of Climate Finance 2017 [R/OL]. https：//climatepolicyinitiative.org/wp-content/uploads/2017/10/2017-Global-Landscape-of-Climate-Finance.pdf.
③ 全球环境基金和绿色气候基金官网.

战。印度为实现"国家自主贡献"（Intended Nationally Determined Contributions，INDC）的目标，需要 2000 亿美元的气候资金①，53 个非洲"国家自主贡献"中，有 43 个国家到 2030 年国家级投资需求将达到 11500 亿美元，23 个国家的减缓投资需求的资金预计 82% 以上来自国际社会。进一步将国家需求分解到各个行业，能源领域是第二大投资需求，占比为 37%（1590 亿美元），而农林和其他土地利用行业为 42%，废物行业占 20%②。

圣乔治地区合作中心（St George Regional Collaboration Center，SGRCC）所做的关于国家自主贡献计划的问卷调查③显示，目前非洲国家亟需能力建设，包括提高政府部门构建可推广和刺激私营部门参与气候融资的法律和政策框架的能力，此外，虽然国际气候基金提供了较多获取资金的机会，但非洲国家普遍面临设计符合气候基金要求的提案能力低下的问题，因而无法充分利用现有的资金机会。④ 东非的关注焦点在于实施具体项目和方案，该地区的一个关键挑战是制订一个包含多样化的清洁能源技术解决方案，通过增强地域可用性、负担能力和可靠性确保整体国家能源安全⑤。埃塞俄比亚和肯尼亚两国的经验也表明地方级别政府缺乏技术专长以支持当地项目的设计和实施，导致拖慢项目的发展，阻碍了国际气候资金在发展中国家的本地化，不利于资金流向最迫切需要的项目领域和地区。亚太地区的交通运输、垃圾处理和农业是最迫切需要支持的领域，以完成长期的国家自主贡献目标。拉丁美洲地域的国家更关注于如何刺激民间资本的投资，以及如何建立 MRV 系统。

2. "一带一路"沿线国家环境承载力异常脆弱，亟需绿色投资

自 2013 年习近平主席提出"一带一路"倡议以来，我国积极提升与沿

---

① 详见：http://www.eurasiareview.com/30082017-climate-finance-in-india-a-case-of-policy-paralysis-analysis/.
② AfDB. African INDCs: Investment needs and emissions reductions in the energy sector [EB/OL]. http://www.vivideconomics.com/wp-content/uploads/2016/11/AfDB-Report-2016.pdf.
③ St George Regional Collaboration Center. Report on the nationally determined contributions survey conducted by the Nairobi Framework Partnership in 2016 [EB/OL]. http://unfccc.int/files/secretariat/regional_collaboration_centres/application/pdf/report_ndc_survey_final.pdf.
④ 详见：http://www.linkedin.com/pulse/africa-needs-climate-finance-capacity-access-funds-critical-domfeh/.
⑤ St George Regional Collaboration Center. Report on the nationally determined contributions survey conducted by the Nairobi Framework Partnership in 2016 [EB/OL].

线各国的经贸合作水平,加强与沿线国家交流对话,推动政策沟通和战略对接,"一带一路"是目前世界上跨度最长、发展潜力最大的经济走廊,涉及欧亚大陆 60 多个国家,主要包括中亚地区和东南亚各国,这些国家普遍存在对外开放程度低、基础设施建设落后和社会经济发展水平低等诸多问题,其中中亚地区多为沙漠,缺乏水资源;东南亚地区热带雨林面积快速缩减;南亚的沿海城市则大量向海洋排污,海洋资源受到严重污染。① 这些沿线国家中,超过一半的国家经济发展模式比较粗放,单位 GDP 能耗和二氧化碳排放超过世界平均水平 1.5 倍②。可见这些国家的环保形势极不乐观,对于绿色投资的需求很大。中国应将此作为重要契机,推动沿线国家的绿色投资,这不仅可以帮助沿线国家转变经济结构实现可持续发展,同时可以促进国内相关产业的技术提升和设备优化,倒逼企业加快研发和激活其创新能力③。

---

① 杨振,申恩威."一带一路"倡议下加快沿线国家绿色投资的探讨 [J]. 2016 (9):21 - 24. 对外经贸实务.

② 中评网. 应对气候变化,"一带一路"的重要内容【N/OL】. http://www.crntt.com/doc/1046/6/6/4/104666452.html? coluid = 7&kindid = 0&docid = 104666452.

③ 杨振,申恩威."一带一路"战略下加快沿线国家绿色投资的探讨 [J]. 2016 (9):21 - 24. 对外经贸实务.

四、特朗普政府对全球气候融资的影响

美国总统特朗普于2017年1月上任后对于气候变化问题采取了与奥巴马截然不同的态度,对美国能源与气候政策进行了大刀阔斧的改革。在他采取的一系列举措中,对全球气候治理、气候投融资影响最深远与广泛的政令包括:取消"气候变化行动",发布"美国第一能源计划",宣布退出《巴黎协定》,大幅度削减气候政策和科研相关预算,限制环保局,重振煤炭行业,解除政府在能源领域的多项监管措施,停止对联合国气候变化项目的资金支持。具体政策内容与发布时间参见表4-1。

表4-1 特朗普政府对全球气候治理和气候投融资有直接影响的系列举措

时间	主要政令与措施
2017年1月21日	发布"美国第一能源计划",取消"气候变化行动"
2017年3月16日	《美国第一财政预算计划》和《推动能源独立和经济增长的总统行政命令》大幅度削减了气候政策和科研相关的预算,停止向绿色气候基金(GCF)以及气候投资基金(CIFs)继续提供资金支持,同时也减少双边气候援助
2017年6月1日	宣布退出《巴黎协定》
2017年6月8日	通过了《金融选择法案》,据此,投资人无法再通过股东提案程序要求企业披露气候风险信息

美国本土大型企业、全球投资者网络、州、市政府、高校与智库组织已针对上述政令采取了自下而上的自主行动。但根据已有一系列模拟研究显示,美国存在无法实现其到2025年在2005年排放量基础上减少26%~28%目标的风险,这又容易引发连锁效应,危及《巴黎协定》的全球目标[①]。另外,美国停止向绿色气候基金(GCF)以及气候投资基金(CIFs)继续提供资金支持,同时减少双边气候援助等举措也将对全球气候融资格局带来深远影响。

---

① Climate Initiative & MIT. Analysis: U. S. Role in the Paris Agreement [N/OL]. https://www.climateinteractive.org/analysis/us-role-in-paris/.

四、特朗普政府对全球气候融资的影响

（一）全球气候资金流的影响

1. 奥巴马时期美国对全球气候融资领域的贡献概况

奥巴马在执政期间通过多边开发银行、美国进出口银行、美国海外私人投资公司和气候基金等多个渠道为发展中国家提供应对气候变化的援助资金。

2010年到2015年，美国提供给发展中国家的气候援助方面的公共资金总额为156亿美元，这些纳入统计的资金包括了双边机构、多边开发银行、发展援助和官方出口信贷机构提供的资金①。

UNFCCC的相关数据显示，美国在2013年和2014年两个财政年度向发展中国家提供的公共资金支持分别为48.35亿美元和51亿美元②，向UNFCCC框架下的气候技术中心和网络贡献了200万美元③，向绿色气候基金承诺了30亿美元，但直到奥巴马卸任美国向该基金的实际拨款额为10亿美元左右。从2013年和2014年两个年度美国国会拨款的地理流向来看，42%用于支持亚洲地区，36%用于支持非洲地区，19%支持拉丁美洲和加勒比地区，剩余部分则用在欧洲和中东的发展中经济体。④

2014年发达国家向发展中国家提供的公共资金总额为432亿美元⑤，其中美国贡献了51亿美元，占全球总值的11.8%。公共资金流动对私人投资者具有催化作用，根据OECD的报告测算，2014年发达国家通过双边和多边

---

① Switzer Foundation. Green Finance: The next frontier of US – China Climate Co – operation [N/OL]. https://www.switzernetwork.org/switzer – fellow – thought – leadership/green – finance – next – frontier – us – china – climate – cooperation.

② U. S. Department of State. 2016 second biennial reports of United States under the UNFCCC [EB/OL]. https://unfccc.int/files/national_reports/biennial_reports_and_iar/submitted_biennial_reports/application/pdf/2016_second_biennial_report_of_the_united_states_.pdf.

③ U. S. Department of State. 2016 second biennial reports of United States under the UNFCCC [EB/OL]. https://unfccc.int/files/national_reports/biennial_reports_and_iar/submitted_biennial_reports/application/pdf/2016_second_biennial_report_of_the_united_states_.pdf.

④ U. S. Department of State. 2016 second biennial reports of United States under the UNFCCC [EB/OL]. https://unfccc.int/files/national_reports/biennial_reports_and_iar/submitted_biennial_reports/application/pdf/2016_second_biennial_report_of_the_united_states_.pdf.

⑤ UNFCCC. 2016 biennial assessment and overview of climate finance flows report [EB/OL]. http://unfccc.int/files/cooperation_and_support/financial_mechanism/standing_committee/application/pdf/2016_ba_summary_and_recommendations.pdf.

机制所撬动的私人资金规模为 167 亿美元①，这意味着每花费 1 美元的公共资金可撬动 0.38 美元私人资本。美国撬动的私人资金具体金额尚无统计数据，但我们假设各个国家撬动私人资金的能力相同的话，可粗略估算美国 2014 年所撬动的私人资金规模约为 19.38 亿美元。根据上述估算方式可粗略得出美国在 2014 年度提供给发展中国家的公共和私人资本总额大约为 70.38 亿美元。

2. 特朗普政府大幅削减气候资金支持

特朗普在 2018 财年预算计划中将环境保护署的预算削减了 31.4%，由环境保护署执行的"清洁电力计划"、国际气候变化项目及气候变化研究与合作项目等不再获得资金支持。美国航天局预算从 193 亿美元减少至 191 亿美元，其中地球科学研究将削减 1.02 亿美元，且几乎完全针对研究气候变化的项目。取消美国国家海洋和大气管理局的海洋管理、研究和教育经费 2.5 亿美元。另外，将美国国际开发署的预算削减 28%。②

除了对美国国内基础科学研究相关经费进行大幅削减外，白宫方面也表示将取消多项海外援助资金，对联合国的出资削减幅度可能高达 50%。

具体到提供给发展中国家的资金支持方面，白宫公布的预算方案显示，多边开发银行的拨款预算将减少 42.1%，发展援助的拨款预算削减 79.6%，美国国际开发预算将削减 27.1%，其他发展及人道主义援助预算将削减 82.7%③。被削减的资金中指向发展中国家应对气候变化的比例不详，但假设美国出资的援助资金按均等比例分配到各个项目中，可粗略预测美国通过多边开发银行、发展援助、美国国际开发署和其他发展及人道主义援助向发展中国家提供的应对气候变化的援助资金将分别减少 42.1%、79.6%、27.1% 和 82.7% 左右。由于特朗普将大幅提升美国国防预算，且对气候变化采取质疑态度，削减的预算中将大比例指向应对气候变化领域，因此以上预

---

① OECD, CPI. Climate Finance in 2013 – 14 and the USD 100 billion goal [EB/OL]. http://www.oecd-ilibrary.org/docserver/download/9715381e.pdf?expires = 1517901239&id = id&accname = guest&checksum = 75A4697E7FA9ADF080FE02BD7169CB4B.

② U.S. Government. Budget of the U.S. Government. A new foundation for American Greatness [EB/OL]. https://www.whitehouse.gov/sites/whitehouse.gov/files/omb/budget/fy2018/budget.pdf.

③ New York Times. Trump Budget Details [N/OL]. https://www.nytimes.com/interactive/2017/05/23/us/politics/trump-budget-details.html?mcubz = 3.

估数据仍属较为保守。此外，特朗普明确宣称将不再向绿色气候基金提供资金，这意味着奥巴马时期承诺的 30 亿美元中的 20 亿美元将失去保障。

要保证联合国气候变化框架公约和全球应对气候变化的承诺不变，由美国削减财政预算所带来的资金缺口需要寻找新的替代途径，这将对全球气候治理格局及中国所扮演的角色带来影响。事实上，美国在全球气候融资体系中所发挥的作用一直有限，而中国对外援助额近几年占全球比重不断提高，目前已经成为世界第四大对外援助国①。尽管欧洲金融部门和气候智库一直在努力重建气候融资体系，撬动私人资本，并且不断进行气候金融产品和工具的创新，但气候融资热点已经转向中国、印度等亚洲地区，全球气候领导力正在进入更迭期。

（二）全球气候治理格局的变化

1. 气候融资领导力正进入更迭期

特朗普宣布退出《巴黎协定》的举动，意味着美国放弃全球气候治理的领导力。同时欧盟正疲于应对英国脱欧及难民潮等一系列问题，在领导气候问题上显得心有余而力不足②，因此当前国际社会对中国填补美国退出《巴黎协定》所留下的领导力真空抱有较高的期望。

美国退出《巴黎协定》很容易让人联想到 2001 年布什总统退出《京都议定书》时发生的连锁效应，当时不少发达国家纷纷效仿或降低减排的执行力度，致使《京都议定书》的执行受阻。但不少专家认为，《巴黎协定》不太可能步《京都议定书》的后尘③。事实上，美国的退出并没有像各方此前所忧虑的重创其他缔约国应对气候变化的决心。相反，在特朗普宣布退出后各国领导人纷纷表示谴责并重申了执行《巴黎协定》的决心。来自国际社会的强烈而迅速的积极回应表明过去 20 年所建立起来的全球气候变化治理体系

---

① 胡鞍钢，张君忆，高宇宁. 对外援助与国家软实力—中国现状与对策 [J]. 2017, 70（3）. 武汉：武汉大学学报，2017.

② 财新网. 美国退出重创《巴黎协定》，中国能否接盘全球领导力真空 [N/OL]. http://china.caixin.com/2017-06-02/101097119.html.

③ 新华社. 特朗普退出《巴黎协定》与小布什退京都议定书有何区别？[N/OL]. http://hunan.voc.com.cn/xhn/article/201706/201706061729545724.html.

的韧性①。2017年7月的汉堡二十国集团峰会上，特朗普在气候议题上提出与各国相反的立场。美国领导力的衰弱不仅表现在气候治理领域，而且在其他国际事务中也正发生改变，如特朗普质疑美国在北约所扮演的角色，退出跨太平洋伙伴关系，大幅削减联合国项目的资金支持等。这将进一步加剧其在全球气候融资方面领导力的下降。

总而言之，美国的退出不会像很多人所担心的那样将动摇全球气候治理机制的根基，反而可能会激化更多积极行动。与此同时，气候领域的领导力正在进入更迭期，气候融资的地理热点逐渐转向中国、印度等亚洲地区，而美国依然保持着新能源技术优势、强大的融资体系与金融市场活力，因此在短期内，气候融资很难再形成单一的核心领导力。

全球气候融资治理格局正悄然发生变化的一个早期信号是发展中国家内部和发展中国家之间所发起的气候倡议。例如东盟和非洲联盟等区域将气候融资纳入其相关举措，发展中国家主导的多边开发银行和基金如新开发银行、亚洲基础设施投资银行和丝绸之路基金等表现出其在气候融资领域的雄心。这些变化表明发展中国家在气候治理中将具备更大的自主权。反映这一变化的最具代表性的则是联合国气候变化框架公约下的气候谈判从《京都议定书》模式转变为《巴黎协定》模式，前者以自上而下的方式作为行动机制，先确定总体排放目标再按名额分配至各个发达国家，发展中国家并不参与；而后者则以自下而上的方式运作，各个国家根据自身的情况制定减排目标，发展中国家具有更多自主权和参与度。

2. 向新能源转型的全球趋势未受明显冲击

新能源领域的投资在过去几年一直在增长。相关数据表明，即使在2008年到2012年的全球经济衰退期，新能源领域的投资增长率仍高于其他领域②。美国政府的新政策很难改变这一趋势，其中一个主要原因是投资者对低碳与新能源领域的信心未受到挫伤。

新能源领域的投资在过去几年迅速崛起，部分得益于清洁能源成本的明

---

① Michele M. Betsill. Trump's Paris withdrawal and the reconfiguration of global climate change governance [J]. Chinese Journal of Population Resources and Environment.

② Joseph P. Tomain. A US Clean Energy Transition and the Trump Administration [J]. Social Science Electronic Publishing.

显下降,例如风电的成本从2009年的每兆瓦101~169美元到2016年每兆瓦32~62美元,下降了66%,太阳能成本从2009年的每兆瓦323~394美元到2016年的49~61美元下降了85%。

全球层面,我们发现,在特朗普宣布退出《巴黎协定》的消息后,VIX波动性(恐慌)指数从10.20下降到9.70,表明退出《巴黎协定》的最终决策平息了猜测。欧元兑美元基本持平。此外,MSCI世界低碳目标指数、Wilder Hill新能源全球创新指数(NEX)自2017年年初以来普遍上涨的势头未受到美国退出《巴黎协定》这一决定的影响;Stowe全球煤炭指数在特朗普宣布退出《巴黎协定》后几天出现小幅下跌,但随后一改2017年4月以来的下跌趋势开始持续走高;标普500石油、天然气勘探和生产行业GICS水平则并未受到特朗普"美国第一能源政策"和退出《巴黎协定》决定的提振,自2017年初到7月底一直延续下跌趋势。

从我国新能源与传统能源板块指数的反应来看,在美国宣布退出后的几天内,清洁能源类股票的表现仍然超越了市场和化石燃料类股票。从2017年走势来看,中国低碳指数从年初(1月3日)的4508.5733至年末(12月29日)的5567.1042,上涨了23.47%,太阳能发电指数从2017年3月开始走跌,但自6月"退出《巴黎协定》"决定宣布后上涨了6.6%,表明决定的宣布平息了市场的猜疑情绪。总体而言,低碳投资者对清洁能源领域的投资信心并未因特朗普发布的一系列"去气候化"政策而受到明显影响。

3. 美国的退出或将挤压中国碳排放空间且增加中国减排成本

已有一系列模拟研究显示,美国存在无法实现其到2025年在2005年排放量基础上减少26%~28%目标的风险。根据戴瀚程等人的最新研究,美国退出《巴黎协定》将会挤压中国的碳排放量空间并增加中国的减排成本。具体而言,在实现2摄氏度温控目标不变的前提下,如果到2025年美国在2005年的排放量基础上减排20%、13%和0%,在这三种不同的情景下,将分别导致中国的碳排放量被挤压1.7%、2.8%和5%,碳价将会上涨4.4~14.6美元/吨,中国的GDP损失到2030年可能将达到219.8亿~711亿美元[①]。从

---

① Dai Han-Cheng, Zhang Hai-Bin, Wang Wen-Tao. The impacts of U.S. withdrawal from the Paris Agreement on the carbon emission space and mitigation cost of China, EU, and Japan under the constraints of the global carbon emission space [J]. 2017, 8 (4): 226-234. Advance in Climate Change Research.

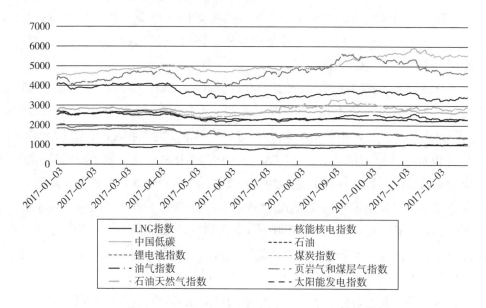

数据来源：Wind 数据库。

图 4-1 中国主要股指自 2017 年 1 月至 12 月的表现

积极层面来看，美国的退出及在应对气候变化问题上的倒退可能进一步加固中国目前在清洁能源领域的领先地位[①]。当前国际社会对中国填补美国退出协定所留下的全球气候领导力真空抱有较高的期望。有学者提出中国必将在全球重新构建共享式领导力，由气候五国 C5 的合作伙伴关系（中国、欧盟、印度、巴西和南非）代替过去的中美 G2 的合作伙伴关系来共同领导全球气候治理问题。[②]

（三）中国在全球气候融资中的新角色

美国在应对气候变化方面的倒退恰逢中国的崛起，一定程度上助力和塑造了中国在全球气候融资格局中的新角色。中国曾被认为是哥本哈根气候谈判失败的主要障碍之一，但又被广泛认为是巴黎谈判取得成功的关键推动力。

---

① Zhang Hai-Bin, Dai Han-Cheng, Hua-Xia Lai, Wang Wen-Tao. US withdraw from Paris [J]. 2017, 1 (6). Advance in Climate Change Research.
② Zhang Hai-Bin, Dai Han-Cheng, Hua-Xia Lai, Wang Wen-Tao. US withdraw from Paris [J]. 2017, 1 (6). Advance in Climate Change Research.

四、特朗普政府对全球气候融资的影响

在短短几年之间,中国在应对气候变化全球合作与博弈的空间角色上发生了明显的变化,这种变化交织着国际和国内层面的多种原因。在国内层面,中国正式将生态文明作为国家发展战略,开始寻求低碳转型,且自从2016年中国人民银行等七部委联合发布《关于构建绿色金融体系的指导意见》后,中国加快了绿色金融发展的进程。在国际层面,中国是《巴黎协定》生效的重要推动者。在气候融资领域,中国也发起和参与了很多融资举措。习近平总书记在十九大报告中指出:"(中国)引导应对气候变化国际合作,成为全球生态文明建设的重要参与者、贡献者、引领者",此外,他进一步强调"要坚持环境友好,合作应对气候变化,保护好人类赖以生存的地球家园"。

在美国退出全球化领导和发达国家反全球化大趋势的背景下,习近平总书记主张共同努力应对气候变化,并且强调了贸易、金融和治理全球化的互惠互利。在中国这样的倡导下,全球应对气候变化工作目前由中国和欧盟共同牵头。欧盟委员会主席在欧盟和中国首脑会议的演讲也更坚定了这一点:"就欧洲方面而言,我们很高兴地看到中国与我们在气候变化问题上的观点与立场一致,这是有帮助的且有责任感的,这也代表中国将和欧盟共同携手推进执行《巴黎协定》的承诺。"①

从气候融资角度来看,中国已经启动并且参与了大量的融资倡议。公共资金方面,主要渠道包括多边开发银行、基金以及中国政策性银行。首先,就多边开发银行而言,主要举措包括设立新开发银行、亚洲基础设施投资银行,并有意设立上海发展合作银行②,这些机构都表现出了对气候领域投资的雄心。其次,在基金方面,中国已经建立了14只国际基金,这些基金都将气候融资领域作为优先投资领域。最后,中国政策性银行以及国有金融企业在国外支持气候相关项目,如国家开发银行和中国进出口银行,自2007年以来已经为国外的水电站项目融资达320亿美元③。

---

① European Commission. Press Release:EU – China Summit:moving forward with our global partnership [N/OL]. http://europa.eu/rapid/press – release_ IP – 17 – 1524_ en. htm.

② Xinhua. Shanghai Cooperation Organization prime ministers meet in Bishkek [N/OL]. http://news. xinhuanet. com/english/2016 – 11/03/c_ 135803540. htm.

③ Financial Times. China poised to lead on green finance at G – 20 meeting [N/OL]. https://www. ft. com/content/65ff1e54 – ec78 – 380e – ac10 – 5a633eea983b.

五、多边开发银行的合作

（一）多边开发银行间的合作潜力巨大

满足基础设施的投融资需求是当前国际金融组织面临的重大挑战之一。据亚洲开发银行预测，截至2030年，亚洲地区每年的基础设施资金需求达1.7万亿美元[①]。2015年在亚的斯亚贝巴召开的联合国国际金融发展会议就如何满足该需求进行讨论，强调了多边开发银行的重要性，并呼吁搭建全球性论坛以加强多边开发银行的合作[②]。会议强调了多边开发银行在融资和知识共享、低波动性和逆周期性信贷、长期贷款、优惠贷款及调动金融市场方面的重要作用，这些都是实现可持续发展目标的重要组成部分[③]。

多边开发银行可以在金融和技术机制、社会和工业部门以及地理区域等多方面展开合作。另外，多边开发银行在金融发展领域的标准化合作模式可以在社会和环境保护以及不同金融工具的使用方面起到示范性作用。近期建立了一些新的多边开发银行，如新开发银行和亚洲基础设施投资银行，以及即将设立的上海发展合作银行，都有中国的参与或引领。这些多边开发银行的成立旨在支持而非改变现有全球金融体系，因此，其运营与传统多边开发银行虽平行，但有交叉领域，为彼此合作提供了空间。

由多边开发银行提供的国际气候资金是将联合国气候大会提出的艰巨任务转化为实际可操作的绿色低碳行动的重要金融工具。多边开发银行的气候融资需求正不断增加，与此同时与气候相关的基金数目也不断增多。在此大

---

① ADB. Meeting Asia's Infrastructure Needs［EB/OL］. https：//www.adb.org/sites/default/files/publication/227496/special‑report‑infrastructure.pdf.

② UN. Addis Ababa Action Agenda［EB/OL］. http：//www.un.org/esa/ffd/wp‑content/uploads/2015/08/AAAA_ Outcome.pdf.

③ UN. Addis Ababa Action Agenda［EB/OL］. http：//www.un.org/esa/ffd/wp‑content/uploads/2015/08/AAAA_ Outcome.pdf.

五、多边开发银行的合作

趋势下，多边开发银行之间的合作潜力巨大，既有共同的组织利益，也有较大的环境影响。通过合作带来的金融资源的规模化效益、技术知识的结合和本土正当性将可大幅增加整体影响力。

本章将主要对亚洲基础设施投资银行与欧洲投资银行的合作模式进行探讨，以期为其他多边开发银行的合作提供范本。由于欧投行希望拓展不同地域的业务，而亚投行的运营能力愈加增强，两家银行的合作具备极大的实操性。同样，由于欧投行愈发关注气候及绿色投资，亚投行成立之初即带有绿色本质，双方对气候与绿色金融领域的关注也提供了明确的合作方向。2016年5月30日，亚投行与欧投行签署了谅解备忘录，并表达了彼此的合作意向，但具体实践方式仍有待商榷。目前双方合作的唯一具体案例是亚投行和世界银行共同出资支持的跨安纳托利亚天然气管道项目，该项目获得了来自亚洲开发银行、欧洲复兴银行和欧洲投资银行等多边开发银行的贷款。亚投行行长金立群曾发表讲话提到双方的合作潜力："该意向书对于拓展我们双方的合作以应对全球巨大的基础设施融资需求至关重要。"欧投行行长Werner Hoyer则在讲话中提道："我们期待与亚投行在应对全球性挑战上的共同合作，尤其是应对气候变化、提供可持续交通及清洁水资源的项目。"①

本报告运用三维分析法来具体分析亚洲基础设施投资银行与欧洲投资银行的各种可能合作领域，再就气候融资及绿色金融层面的合作可能性展开深入探讨。整体合作可能性的概要分析是衡量气候融资与绿色金融合作可能性的先决条件。分析将从授权文件、运营能力、政治环境三个维度展开，如图5-1所示，多边开发银行可在三个方面开展合作：左栏为所有可能的金融、技术和联合工具；中间栏为所有可能合作的行业，根据欧投行、世界银行及海外发展机构对项目的归类标准进行分类。这些行业虽有重叠，但其划分方法囊括了几乎所有标准；右栏列明了可能合作的地域和国家，任何多边开发银行之间的合作必须符合三个方面中的至少一条，比方说对巴基斯坦（地域）能源领域（行业）的咨询服务（工具）。

---

① China Daily. AIIB, EIB sign a pact to help finance infrastructure [EB/OL]. http://www.chinadaily.com.cn/business/2016-05-31/content_25544988.htm.

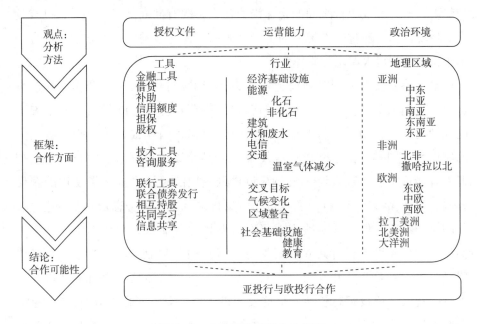

图 5-1 多边开发银行合作框架

（二）亚投行与欧投行的合作

1. 亚洲基础设施投资银行

亚洲基础设施投资银行（Asian Infrastructure Investment Bank，AIIB）由中国倡议设立，主要由亚洲及欧洲的 50 个成员国组成，初始筹资规模 1000 亿美元。作为发展中国家主导的大型多边银行，其设立宗旨是在现有的全球金融体制下运行，并以精益、绿色、廉洁为目标。亚洲成员国占 78% 的投票权，虽然没有正式否决权，但任何一项决议的通过需要 75% 以上的成员同意，因此中国 27% 的投票权构成了其领导地位及实际意义上的一票否决权。随着新成员的加入，中国的投票权会相应下降，如亚投行行长金立群所表达的，该机构作为一家民主机构的特性将更加突出[1]。与其他多边开发银行相似，亚投行有望在国际金融市场上发行债券为项目提供融资支持。虽然亚投行已在 2017 年 6 月获得穆迪 AAA 评级，但由于其充足的初始资本及采取的

---

[1] China Daily. AIIB Chief rules out China vetopower [N/OL]. http：//www.chinadaily.com.cn/business/2016 - 01/27/content_ 23265846. htm.

逐渐扩张的投资规模策略，导致目前尚未在国际金融市场发债。亚投行的三个优先目标包括：通过对可持续基础设施的融资支持发展中国家实现其环境及发展目标；促进亚洲地区的互联互通和经济一体化；加强与其他多边开发银行、政府及私有金融机构合作来撬动社会资本。

2017 年 6 月发布的亚投行能源策略表明，其目标是为亚洲数百万人口提供电力系统的融资支持，并促进亚洲国家实现其环境目标[1]，亚投行将通过投资可再生能源和能源效率项目、维护与升级现有电力系统、搭建网络来实现此目标。该策略特别强调将与其他多边开发银行在这些领域展开合作。虽然绿色概念是其宗旨的核心，但对于贴标为绿色融资的金额却未作具体量化阐述。

2. 欧洲投资银行

欧洲投资银行（European Investment Bank，EIB）是欧洲经济共同体成员合资经营的金融机构，于 1958 年根据《罗马条约》的规定成立，其经营不以营利为目的。该行股本金为 2430 亿欧元，是世界上最大的多边开发银行，主要代表欧盟成员国的利益并对共同体内外有利欧盟成员国的可持续发展项目提供融资支持。欧投行将其优先融资的项目定位在基础设施、创新、气候变化、减缓与适应行动。欧投行由欧共体委员会管理，其组织及治理机构在运营中扮演了重要角色。作为欧盟的机构，欧投行的战略方针与其他欧盟机构一致，并与其他机构紧密合作互相支持[2]。

为了最大化其贷款，欧投行从国际资本市场获得资金，大部分资金来自于国际资本市场上的债券发行，其债券的主要购买者包括来自欧盟和其他国家的机构投资者、国际市场上的普通投资者[3]。该行主要通过信贷、混合及咨询等模式开展业务。其中，信贷占该行总融资规模的 90%。对于大规模项目（融资需求超过 3000 万美金的项目），欧投行可资助项目总体资金量的 50%，但平均来看资助额一般占总资金需求的 30%[4]。对国际金融机构而言，

---

[1] ADB. Meeting Asia's Infrastructure Needs [EB/OL]. https://www.adb.org/sites/default/files/publication/227496/special-report-infrastructure.pdf.

[2] EIB. Part of the EU family [EB/OL]. http://www.eib.org/about/eu-family/index.htm.

[3] EIB. Overview of EIB [EB/OL]. http://www.eib.org/investor_relations/overview/index.htm.

[4] EIB. The European Investment Bank at a glance [EB/OL]. http://www.eib.org/infocentre/publications/all/the-eib-at-a-glance.htm.

信贷是一种较为普遍的融资方式。混合，在欧投行自己的界定中是指融合了多种复杂金融工具的模式。咨询，是欧投行的另一种业务模式，作为大型投资机构，欧投行无疑是金融领域的专家，因而可以为其他利益相关者提供咨询建议，例如，为如何执行项目和找到融资途径提供建议，或为如何在国家层面制定监管框架以刺激投资提供政策建议。①

近些年来，欧投行正逐步扩大其对气候与绿色金融领域的关注，对低碳及气候韧性相关项目的融资金额已达到其总投资额的25%。其中，信贷总额的10%是发放给非欧盟国家的，而这部分资金中35%运用于绿色项目。此外，欧投行在2007年发行了全球第一支绿色债券，是绿色金融领域当之无愧的先行者。欧投行的气候战略能清晰地反映其对绿色金融的态度："为了调动资本市场，实现全球控制气温升高在2摄氏度内并适应气候变化带来的影响的目标，欧投行应在金融机构中发挥主导作用。"②

3. 亚投行与欧投行的合作潜力

亚投行与欧投行的合作潜力如图5-2所示。方块表示具有较强的合作可能性，圆形表明没有强有力的因素增强或抑制合作，而三角形代表有明确阻碍合作的因素。

通过对亚投行和欧投行的授权文件的分析，可见两家银行的使命基本趋同。虽然贷款、担保和知识共享是两个机构明确的优先合作领域，但在其他机制、行业或地理区域也有很大的发挥空间。亚投行在创立初期需要稳健的资产负债表，同时因实践经验相对不足，限制了其融资工具的范围和行业、技术以及地理区域的风险偏好。从运营能力上看，欧投行作为经验丰富的全球性多边开发银行，其技术权威性能够与亚投行的本土正当性形成互补作用。就政治动力而言，欧盟重点投资相邻地区的能源和交通基础设施。欧盟希望通过促进两家银行合作，推动拓展欧投行在亚洲的各项业务。而亚投行与其他多边开发银行（包括欧投行）合作能够降低风险，合作范围包括所有机制、"一带一路"倡议强调的多数行业以及幅员辽阔的中国周边国家。亚洲发展中国家作为亚投行成员和初期项目接收者，更关注多样性的基础设施融

---

① EIB. Advising [EB/OL]. http://www.eib.org/products/%20advising/index.htm.
② EIB. EIB Climate Strategy [EB/OL]. http://www.eib.org/infocentre/publications/all/eib-climate-strategy.htm.

五、多边开发银行的合作

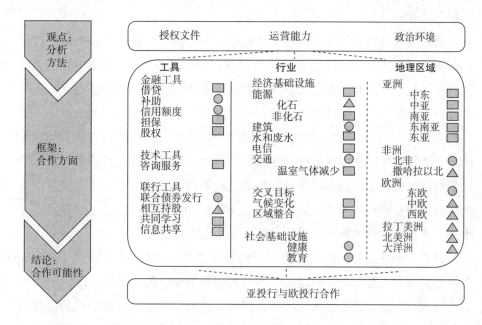

图 5-2　亚投行与欧投行的合作模式

资,因而会更支持多边开发银行的合作。另外,由于大量的地域性和全球性支持确保了亚投行的合法地位,非亚投行成员(包括美国和日本)已无法对亚投行与欧投行的任何合作框架构成实质性限制。因此从总体来看,两家新旧多边银行完全有可能在平等合作的前提下实现共赢。

4. 在气候融资与绿色金融方面的合作潜力

在对亚投行和欧投行合作潜力分析的基础上,可以深入探究双方在气候融资与绿色金融领域具体合作的可能性。首先,在工具方面,对组织授权、运营能力和政治动力分析表明,贷款、担保和知识共享是两者合作的优先领域。对于气候融资与绿色金融而言以上论点依然成立。由于欧投行一直是绿色金融方面的创新者,因而在金融工具方面具备的技术专长对亚投行有极大的帮助。欧投行与亚投行合作发行绿色债券的可能性是最受到关注的,全球绿色债券市场在过去几年呈指数级增长,多边开发银行已成为这个趋势的重要发起人和贡献者,如亚洲开发银行在 2017 年 8 月 2 日发行了 12.5 亿美元的绿色债券①。

欧投行可以根据其自身经验,通过最简单的银行间知识转移协助亚投行发起首只绿色债券,或联合发行绿色债券,如"亚洲绿色基建债券"。而亚投行可依靠欧投行的技术专长来发展绿色债券市场,以便积累经验在将来单独发行绿色债券。截至目前,亚投行尚未开始在全球金融市场筹集资金。亚投行有着1000亿美元的初始资金。随着全球资本市场对绿色债券需求的日益增加,亚投行可以利用这个机会在绿色债券市场建立自己的影响力。而对于欧投行的益处在于能利用联合发行增加其在亚洲的投资规模,并完成其作出的将对非欧盟成员国绿色贷款增加至总投资额35%的承诺,绿色债券方面开展的合作将同时传递欧盟—亚洲对促进绿色金融发展的决心。

其中一个需要考量的关键因素是两家多边开发银行对绿色的不同定义。尽管双方都将可再生能源和能源效率列为优先领域,但对待化石燃料的态度仍有出入。在欧盟的投资中,欧投行排除了所有可能增加排放的项目,接受所有能源效率项目,包括交通运输项目。相对于建设高速公路这类会导致汽车和卡车的排放量增加的项目,欧投行更倾向于资助铁路或轻轨等公共交通工具,因为此类项目能以较低的排放率实现同样的目的。而亚投行的能源战略中指明投资的项目需要与一个国家的过渡发展时期相适应,使用商业可行的碳排放量最小的技术;且如果可以替代现有低效的系统,则将考虑低碳石油和燃煤电,或者其使用对系统的可靠性和完整性至关重要②。这种界定和标准之间的差异很常见,发达国家将气候变化作为绿色发展的重点,因此大多数情况下排除了所有化石燃料,而发展中国家的绿色概念超越了气候变化。因此,绿色金融合作受限于欧投行较窄的绿色金融定义。

另外需要考虑的是绿色领域的许多技术提供方源于发达国家,尤其是欧盟。因此,在代表欧盟方面,欧投行将有兴趣在最大程度上与亚投行在绿色金融领域合作,以达到为欧洲企业开辟提供通往亚洲国家门户的目的。与其他行业领域的合作相比,绿色投资通常具有更大的技术含量,因而相比其他

---

① ADB. ADB Sells Dual – Tranche $750 Million 5 – Year and $500 Million 10 – Year [EB/OL]. https://www.adb.org/news/adb – dual – tranche – global – green – bonds – spur – climate – financing.

② ADB. Meeting Asia's Infrastructure Needs [EB/OL]. https://www.adb.org/sites/default/files/publication/227496/special – report – infrastructure.pdf.

行业，欧投行将更多地推动这一领域。但是，绿色投资往往风险较高，例如使用较不成熟的技术或频繁变化的政治环境将导致较高的风险。由于亚投行现阶段试图建立较为稳健的财务表现，因此高风险项目对其不具有太高吸引力，但很明显与欧投行合作将有利于亚投行在绿色领域降低投资风险。

关于绿色金融合作的地理范围方面，欧投行在全球范围内运营，而亚投行虽然已对南非、埃及和巴西其他发展中国家表示了兴趣[1]，但其侧重点主要还是在亚洲。亚投行与欧投行在绿色金融领域具有合作潜力的国家包括一些在开发可再生能源领域态度较为积极的亚洲国家，或能源需求增长快速的区域，如巴基斯坦、印度、印度尼西亚等国家。此外，由于欧投行在其他国家的投资有限，在地域上与更临近欧洲的亚洲国家显示出较大的合作潜力，如阿塞拜疆和土耳其。

---

[1] Institute of Development Studies. The Asian Infrastructure Investment Bank: What Can It Learn From, and Perhaps Teach To, the Multilateral Development Banks? [R/OL]. http://www.ids.ac.uk/publication/the-asian-infrastructure-investment-bank-what-can-it-learn-from-and-perhaps-teach-to-the-multilateral-development-banks.

六、政策建议

（一）坚持履行《巴黎协定》，推动全球建立一体化的气候公共物品供给机制

要实现《巴黎协定》设定的目标，中国所担任的角色举足轻重。在美国因为政府更迭导致气候政策转向的大背景下，中国坚持承担在气候治理方面的大国责任在气候变化问题上享有主导权，将成为掌握未来世界主导权的必要条件之一。我国气候治理的引领者需积极推动全球建立一体化的气候公共物品供给机制，成为建立全球新规则制定者和主导者。

目前气候融资体系存在的过度依赖ODA资金体系与融资工具的情况，其根本原因在于全球尚未建立起一个完善的气候公共物品供给理念及融资体系。加速提升气候资金供给总量、效率与效果的前提是为全球减缓和适应战略形成独立的融资机制和相匹配的融资工具，尤其需要重申全球公共财政的核心作用，加强赠款资源的供给和有效利用，以撬动、引领及推动更大规模的私人投资来解决气候公共物品供给不足的问题，同时，也要避免以牺牲国家发展为代价来提供全球公共物品。核心的举措和重点领域包括：（1）充分发挥我国参与和主导的多边平台与机构的全球影响力，推动国际社会、多边发展伙伴关系来重塑国际合作的双轨机制，建立与ODA并行且协同的气候公共物品供给机制，融资体系和工具箱。（2）主张各国政府在充分尊重"共同但有区别的责任"前提下为全球公共物品融资机制贡献公共资金部分作出贡献，并将出资责任贯彻到国家相关部门（环境、能源、金融、卫生或贸易等）的预算框架之中。（3）成立更具包容性的国际公共物品融资机构，承担类似OECD发展援助委员会的职责。（4）在ODA与气候融资双轨模式下推动多边机构的规章制度、部门设置、资金安排与融资工具的创新和发展。

（二）推动新兴多边机构在应对气候变化领域的渠道作用和多方面创新

中国在全球气候治理中的作用愈加重要，其推动设立的新兴多边机构是全球气候治理"软实力"的集中体现，其长期战略目标高于各组织在短期内的经贸利益，其定位和功能需要在原有援助、投资和贸易活动的基础上，从而进一步与我国在全球公共物品提供框架中的引领作用相匹配。建议在未来的发展中通过统一的顶层设计，推动新兴多边机构从治理结构到融资工具的多方面创新。主要的建议包括：

1. 推动符合气候韧性标准的基础设施建设。"一带一路"国家、南方国家均是全球气候风险较高地区，其经济社会最容易受到气候变化威胁，新兴多边机构应将促进对低碳、气候韧性基础设施领域的知识能力建设、标准建设、商业模式创新及撬动私人投资作为其重点领域，建立专门的专家委员会及部门，并设立相应的绩效考核标准，加大资金和人才投入。

2. 在宏观政策层面，各国保证居民生活质量和生存环境的需求与可持续发展目标是相互关联的，因此新兴多边机构应充分考虑受援国的气候目标与政策，参考《巴黎协定》所达成的协议，并结合各国所提出的"国家自主贡献"，保证项目目标符合各方的共同预期。而在具体操作层面，新兴多边机构可以利用创新的交易结构、低风险的技术、本土化的市场专业知识，将私人投资引入低碳基建和其他绿色项目中。

3. 强调与传统多边机构的合作和互补。新兴多边机构与传统多边机构虽然存在竞争关系，但应更强调合作和互补。发展中国家在基础设施发展方面仍面临巨大的资金缺口，需要与各多边机构通力合作，填补缺口。新兴经济体仍然积极支持传统多边机构的发展和改革，同时，它们也将更多地在新兴机构中寻找机会。未来，传统和新兴多边机构可以通过共同资助特定项目以及相互学习彼此的想法和经验来开展合作，从而建立起一个新的国际发展融资系统，以更好地满足发展中国家的需要。

4. 鼓励混合化的融资方式。在发展中国家，公共资金存在较大空缺，而私人资金的流入常常也受到阻碍，因此有必要通过新兴多边机构扩大公共和私人资本的引进与融合。同时，多边与本土发展机构的融资混合也势在必行。

本土发展银行对本地的项目和地方的法规更为了解，而国际或地区组织对于最佳的工程、设计及金融实践有更深刻的认知。因此，混合融资一方面可以扩大融资规模，另一方面也可以提高资金的使用效率，从而构建起良性的系统。

5. 重视本土因素。一些传统多边机构曾把发达国家作为国际发展的核心参与者，并试图将发达国家的行动模式当成帮助发展中国家前进的模板。然而事实上，发展中国家有能力在国际发展中扮演更重要的角色，并且每一个发展中国家都有其特殊的国情，鲜有模板化的路径可以依照。解决发展中国家本土化的问题是新兴多边机构的初衷，这种理念也必须结合到气候融资中去。在气候变化中，发展中国家相比发达国家更为脆弱，但也具有独特的力量。发展中国家需要充分沟通、相互学习，并且结合具体国情找出问题、提出解决方案，从而更有效地整合资源、开展业务。

（三）以气候变化南南合作为突破口，加强我国气候融资软实力

中国气候变化南南合作基金的开启，推动了气候变化南南合作的进一步发展。虽然在对外援助、投资与对外贸易领域缺乏长期的一体化规划，也并没有形成独立的国际发展援助管理体系及管理结构，但我国在应对气候变化、区域协作、绿色金融国际合作等领域取得了一定的国际合作经验，在节能减排的政策实践、方法和技术上都有一定的积累与优势，已经具备开展气候融资软实力建设的基础和条件。据此建议：

1. 发展南南合作的管理体系与成效评估机制。以发达国家为代表的对外援助国都建立了不同程度的管理框架体系，从较为成功的经验来看，这一管理框架主要包括法律和政策基础、组织机构、管理模式、协作模式四个要点。应对气候变化南南合作是中国对外援助中相对较新，并且专业性比较强的领域，亟需建立能够有效运行的管理体系，保证南南合作的顺利实施。另外，为保证实施效果，需要建立和完善成效评估机制，采取宏观评估和微观评估共进的模式。宏观评估主要是总结在一段时期内南南合作的整体效果，而微观评估主要是针对具体项目展开。良好的成效评估有利于项目设计与管理、经验总结与反馈，并有利于提高透明度与公信力。

2. 建立多元化的气候投融资机制。中国已在 2015 年 9 月宣布出资 200 亿

元人民币建立"中国气候变化南南合作基金",用于支援其他发展中国家应对气候变化。南南合作基金的资金来源应不限于无偿捐赠,还需吸引多元化的资金进入,借助专业型的基金管理模式,提高资金的使用效率和效果。由于吸引多元化资金进入需要"有利可图",南南合作基金可以考虑进行一定的"绿色投资",由包括私人资本在内的多元渠道资金共同投资、共担风险,私人资本可获得一定比例的回报或者其他间接回报,基金可以将剩余回报纳入再援助或投资中。另外,中国政府和受援国政府还可以采取税收减免或者其他优惠政策来吸引私人资本加入。南南基金可以作为母基金以多种方式注资其他区域的气候基金,汲取全球经验的同时,谋求与世界银行等传统多边金融机构及亚投行等新兴多边金融机构的协作,在区域气候公共物品提供的政策与实践上取得突破。

3. 加强顶层设计能力的经验输出,探索创新融资工具平台。中国近些年的气候融资体系不断完善,已经积累了较多的发展经验,未来可以重点提供绿色金融发展的经验和能力建设援助。例如在南南气候变化合作基金框架下建设学习网络,推广可刺激私营部门资金流入的政策框架设计经验。此外,为了更有效的利用援助资金,可以借鉴国际经验,设立创新融资工具平台,加大对外气候援助资金的杠杆作用,如大部分国家对于购买主权保险的需求旺盛,但其参与度受到低价或免费保险的可获得性影响很大,中国可以创建提高南南国家气候韧性的计划,为一些由于保费过高而无法购买气候保险的国家提供补助金以降低保费。此外,目前很多发展中国家的金融市场发展仍非常不成熟,各个国家的挑战与在撬动私人资金方面面临的障碍不尽相同,有必要针对当地需求进行调研,以明确需要克服的具体障碍并根据各个国家的国情制订有针对性的方案。同时,可以帮助发展中国家开发更多的金融工具,如巨灾债券,目前比较成功的巨灾债券有墨西哥的地震巨灾债券,可以借鉴地震巨灾债券的经验,并加以推广到极端天气中的应用。此外,也可以通过损失分担等方式来鼓励其他发展中受援国的银行提供绿色贷款,为气候融资降低风险级别,促进资金流动。

4. 充分利用"一带一路"倡议,提高对沿线国家可持续基础设施投资。中国2016年全面落实"一带一路"倡议,进一步为气候变化南南合作搭建了互利共赢的新平台。未来可借助"一带一路"倡议框架,支持其他发展中国家，尤其是最不发达国家和脆弱国家提高适应气候变化的能力，探索具有气候韧性的可持续发展道路，加强对"一带一路"沿线国家可持续基础设施建设的资金支持。基础设施作为经济增长的基础和动力，其互联互通是"一带一路"建设的优先领域。但是"一带一路"沿线众多的新兴经济体国家粗放的经济发展方式和脆弱的自然生态环境加剧了这些国家的气候脆弱性。因此，中国在支持沿线国家的基础设施建设时，应该充分考虑气候变化的影响，把基础设施设计、投资和运营的全生命周期纳入气候风险因素的考量，以国际最佳适应实践为参考，支持具有气候韧性的基础设施建设。此外，大力支持"一带一路"沿线国家适应气候变化的能力建设。气候适应不仅包括基础设施建设等具体的工程项目，更多的是加强气候风险管理。

5. 完善气候适应资金的 MRV 体系建立（Monitoring, Reporting and Verification, MRV）。目前有关气候适应资金的数据大多基于国际公共适应资金，还缺乏对私人适应资金以及国家和区域层面的公共适应资金的 MRV。气候适应活动大多是出于发展中国家的国内需求，是国家和区域层面的行动，很大一部分由预算资金支持[1]。虽然已有国家对这部分资金进行测算，但仍旧只占少数，且其使用的方法各异，缺乏全面性和可比性。未来应当加强对我国及受援国适应资金的监测、核查与报告。

---

[1] Schealatek L, Nakhooda S, Watson C. The Green Climate Fund [EB/OL]. 2016 [2017 - 09 - 10]. https: //www.odi.org/sites/odi.org.uk/files/resource - documents/11050.pdf.

2017 China Climate Financing Report

中央财经大学绿色金融国际研究院
INTERNATIONAL INSTITUTE OF GREEN FINANCE, CUFE

中央财经大学
气候与能源金融研究中心
RESEARCH CENTER FOR CLIMATE AND ENERGY FINANCE, CUFE

About the Report

International Institute of Green Finance, CUFE (IIGF)

International Institute of Green Finance (IIGF) of Central University of Finance and Economics (CUFE), is known as the first international research institute in China whose goal is to promote the development of green finance. The IIGF grew out of the Research Center for Climate and Energy Finance, which was founded in September 2011. IIGF is one of the executive member of Green Finance Committee (GFC) of China Society of Finance and Banking while it has built an academic relationship with the Ministry of Finance. The IIGF aims to cultivate the economic environment and social atmosphere with the spirit of green finance and to build the domestic first – class, the world's leading financial think tank with Chinese Characteristics.

Research Center for Climate and Energy Finance, CUFE (RCCEF)

Founded in September 2011, RCCEF has issued China Climate Financing Report for seven consecutive years till 2017. Based on generalized concept of global climate financing, RCCEF has established an analytical framework of climate financing flow and created a model of China's climate – related financing demands. Besides, RCCEF does in – depth analysis on international climate capital governance as well as China's climate financing development on a yearly basis and has accumulated a series of research findings in this area.

Instructor

WANG Yao — Professor, Director General, International Institute of Green Finance, CUFE
Deputy Secretary General, Green Finance Committee of China Society for Finance and Banking

Authors

LIU Qian	Professor, Research Center for Climate and Energy Finance, CUFE
CUI Ying	Senior Researcher, Research Center for Climate and Energy Finance, CUFE
Mathias Lund Larsen	Researcher, Research Center for Climate and Energy Finance, CUFE
XU Yinshuo	Director Assistant, Research Center for Climate and Energy Finance, CUFE
LUO Tanxiaosi	Researcher, Research Center for Climate and Energy Finance, CUFE
YAO Yingzhi	Assistant Researcher, Research Center for Climate and Energy Finance, CUFE
LOU Huiyuan	Research Assistant, Research Center for Climate and Energy Finance, CUFE
ZHANG Donghou	Research Assistant, Research Center for Climate and Energy Finance, CUFE
XU Lei	Research Assistant, Research Center for Climate and Energy Finance, CUFE

Foreword

The *Paris Agreement* signed in 2015 has entered the stage of implementation. This past year has witnessed a series of profound changes in the global political and economic, which also impacted the climate governance structure and climate financing system. The most representative one of these events was United States President Trump's announcement of the withdrawal from the *Paris Agreement* on 1 June 2017. In spite of this, the determination of all countries to jointly develop green finance and work together to combat climate change has not been shaken. Emerging economies' responding to climate change, in particular, have been increasingly active during the past few years. South – South cooperation is one of those efforts and has been made a priority by emerging economies. Out of all the developing countries' efforts, China's stands out distinctively.

In recent years, China has attached great importance to the development of an ecological civilization, environmental protection, development of the low – carbon economy, and combating climate change. Green finance has become a key concept in China's national policy. During the opening session of the 19th Communist Party of China (CPC) National Congress, General Secretary Xi Jinping stated that "we should persist in the basic national policy of saving resources, protect the environment, and treat the environment as our own life." The term "green" was repeatedly and increasingly mentioned, from just once at the 18th CPC National Congress to 15 times this year, demonstrating China's firm determination to follow the path of sustainable development. Internationally, China also actively advocates for the global development of green finance. Under China's presidency of the G20, green finance became a key theme at the G20 agenda for the first time. In addition, China also serves as a promoter for all countries that signed the *Paris Agreement* and provides climate finance to other developing nations in terms of tackling climate change. Undoubtedly, China is gradually becoming a leader on the stage of international climate governance.

Climate finance has long been a key element of global climate negotiations and an essential tool to address climate change. At present, the global attention is on how to fill the enormous gap between the supply and demand for climate finance. The Research Center for Climate and Energy Finance of the Central University of Finance and Economics (RCCEF) has closely followed the development of international climate finance since 2011. Based on experience and methodologies accumulated over the past seven years, this report reviews the latest developments of 2017, reflects on the existing climate financing architecture, and analyzes how to mobilize private capital and increase the supply of finance for adaptation. In doing so, we hope to contribute wisdom from developing countries to the international community and to offer valueable advice and suggestions to policymakers.

英文缩写索引表

英文简写	英文全称	中文对应名称
A		
AAU	Assigned Amount Units	分配数量单位
ABS	Asset-Backed Security	资产支持证券
ACCF	Africa Climate Change Fund	非洲气候变化基金
ASAP	Adaptation for Smallholder Agriculture Programme	小型农业适应计划
ADB	Asian Development Bank	亚洲开发银行
AF	Adaptation Fund	适应基金
AfDB	African Development Bank	非洲开发银行
AIIB	Asian Infrastructure Investment Bank	亚洲基础设施投资银行
ARC	Africa Risk Capacity	非洲风险防范能力机制
B		
BFIs	Bilateral Financial Institutions	双边金融机构
C		
CBFF	Congo Basin Forest Fund	刚果盆地森林基金
CCCFs	County Climate Change Funds	县级气候变化基金
CCCG	Climate Change Coordination Group	气候变化协调小组
CCER	China Certified Emission Reduction	中国核证自愿减排量
CCF	Climate Change Fund	气候变化基金
CCRIF	Caribbean Catastrophe Risk Insurance Facility	加勒比巨灾风险保险基金
CCWG	Climate Change Working Group	气候变化工作组
CCXG	Climate Change Expert Group	气候变化专家组
CDKN	Climate and Development Knowledge Network	气候与发展知识网络
CDM	Clean Development Mechanism	清洁发展机制
CDMF	Clean Development Mechanism Fund	清洁发展机制基金
CER	Certification Emission Reduction	核证减排量
CFIF	Climate Finance Innovation Facility	气候融资创新机制

续表

英文简写	英文全称	中文对应名称
CIDA	Canadian International Development Agency	加拿大国际开发署
CIFs	Climate Investment Funds	气候投资基金
CTF	Clean Technology Fund	清洁技术基金
COP	Conferences of the Parties	联合国气候变化框架公约缔约方大会
CPI	Climate Policy Initiative	气候政策倡议
CRGE	Climate Resilient Green Economy	气候韧性绿色经济机制
D		
DFIs	Developmental Finance Institutions	发展性金融机构
DECC	Department of Energy and Climate Change	能源与气候变化司
DFAT	Department of Foreign Affairs and Trade	外交与贸易部
DFID	Department for International Development	国际发展部
E		
EBRD	European Bank for Reconstruction and Development	欧洲复兴开发银行
EC	European Commission	欧盟委员会
XCF	Extreme Climate Facility	极端天气防范机制
EIB	European Investment Bank	欧洲投资银行
ERU	Emission Reduction Unit	排放减量单位
F		
FAO	Food and Agriculture Organization	联合国粮食和农业组织
FCPF	The Forest Carbon Partnership Facility	森林碳伙伴基金
FDI	Foreign Direct Investment	外商直接投资
FFEM	French Global Environment Facility	法国全球环境基金
FIP	Forest Investment Program	森林投资项目
FoF	Fund of Funds	母基金
FTT	Financial Transaction Tax	金融交易税
G		
GCF	Green Climate Fund	绿色气候基金
GDP	Gross Domestic Product	国内生产总值
GEF	Global Environmental Facility	环球环境基金
GLO	Global Legislators Organization	全球立法组织

续表

英文简写	英文全称	中文对应名称
GIL	Global Innovation Lab	全球创新实验室
GIZ	Deutsche Gesellschaft für Internationale Zusammenarbeit	德国国际合作机构
GRICCE	Grantham Research Institute on Climate Change and the Environment	格兰瑟姆气候变化与环境研究所
I		
IADB	Inter–American Development Bank	泛美开发银行
IBRD	International Bank for Reconstruction and Development	国际复兴开发银行
ICI	International Climate Initiative	国际气候倡议
IDA	International Development Association	国际开发协会
IEA	International Energy Agency	国际能源署
IET	International Emission Trading	排放贸易机制
IFAD	International Fund for Agricultural Development	国际农业发展基金
IFC	International Finance Corporation	国际金融公司
INDC	Intended Nationally Determined Contributions	国家自主贡献
IPCC	Intergovernmental Panel on Climate Change	政府间气候变化委员会
J		
JFSF	Japan Fast Start Fund	日本快速启动基金
JI	Joint Implementation	联合履约机制
JICA	Japan International Cooperation Agency	日本国际合作署
K		
KfW	Kreditanstalt für Wiederaufbau	德国复兴信贷银行
L		
LDCF	Least Developed Countries Fund	最不发达国家基金
LDCs	Least Developed Countries	最不发达国家
M		
MDBs	Multilateral Development Banks	多边开发银行
MFIs	Multilateral Financial Institutions	多边金融机构
MIGA	Multinational Investment Guarantee Agency	多边投资担保机构
MP	Montreal Protocol	蒙特利尔议定书
MRV	Monitoring, Reporting and Verification	监测、报告和核证体系

续表

英文简写	英文全称	中文对应名称
N		
NAMAs	Nationally Appropriate Mitigation Actions	国家适当减缓行动
NAPAs	National Adaptation Plans of Action	国家适应行动计划
NAPs	National Adaptation Plans	国家适应计划
NCF	National Climate Fund	国家气候基金
NDBs	National Development Bank	国家开发银行
NORAD	Norwegian Agency for International Development Cooperation	挪威国际发展合作署
O		
ODA	Official Development Assistance	官方开发援助
ODI	Overseas Development Institute	海外发展机构
OE	Outbreak & Epidemic	传染性疾病暴发防范机制
OPIC	Overseas Private Investment Corporation	海外私人投资公司
OECD	Organization for Economic Co-operation and Development	经济合作与发展组织
P		
PPCR	Pilot Program for Climate Resilience	气候韧性试点项目
PMR	Partnership for Market Readiness	市场准备伙伴关系
PPP	Public-Private Partnership	政府与社会资本合作模式
R		
RC	Replica Coverage	复制性覆盖
REDD+	Reducing Emissions from Deforestation and Degradation	减少毁林和森林退化造成的温室气体排放
S		
SCCF	Special Climate Change Fund	气候变化特别基金
SCF	Strategic Climate Fund	战略气候基金
SGRCC	St George Regional Collaboration Center	圣乔治地区合作中心
SIDA	Swedish International Development Cooperation Agency	瑞典国际发展合作署
SPV	Special Purpose Vehicle	特殊目的实体
U		
UNDP	United Nations Development Programme	联合国开发计划署

续表

英文简写	英文全称	中文对应名称
UNEP	United Nations Environment Programme	联合国环境规划署
UNFCCC	United Nations Framework Convention on Climate Change	联合国气候变化框架公约
UNISDR	United Nations International Strategy for Disaster Reduction	联合国减灾署
UN–REDD	United Nations Reducing Emissions from Deforestation and Forest Degradation	减少发展中国家毁林和森林退化所致排放量联合国合作方案
USAID	United States Agency for International Development	美国国际开发署
W		
WB	World Bank	世界银行
WBG	World Bank Group	世界银行集团
WFF	Water Financing Facility	水资源融资机制
WMO	World Meteorological Organization	世界气象组织
WRI	the World Resource Institution	世界资源研究所

CONTENTS

I. **Progress in Global Climate Financing** ⋯⋯⋯⋯⋯⋯⋯⋯⋯⋯⋯⋯ 71
   (Ⅰ) Continuous Expansion of Global Climate Financing Gap ⋯⋯⋯⋯ 71
   (Ⅱ) Integrated Capital Supply Mechanism on Global Climate
       Public Goods Not Formed ⋯⋯⋯⋯⋯⋯⋯⋯⋯⋯⋯⋯⋯⋯⋯⋯⋯ 72
   (Ⅲ) Synergy of Different Finance Mechanisms Needs to be
       Improved ⋯⋯⋯⋯⋯⋯⋯⋯⋯⋯⋯⋯⋯⋯⋯⋯⋯⋯⋯⋯⋯⋯⋯⋯⋯ 76
   (Ⅳ) The New and Old Multilateral Development Institutions Need to
       Seek Synergistic Mechanism and Model Innovation ⋯⋯⋯⋯⋯⋯ 77

II. **Progress of Climate Financing in China** ⋯⋯⋯⋯⋯⋯⋯⋯⋯⋯ 84
   (Ⅰ) Gradual Improvement of Climate Change Policy System ⋯⋯⋯⋯ 84
   (Ⅱ) Further Deepening the Adoption of PPP Mode in Green and
       Low–Carbon Fields ⋯⋯⋯⋯⋯⋯⋯⋯⋯⋯⋯⋯⋯⋯⋯⋯⋯⋯⋯⋯ 92
   (Ⅲ) Constantly Strengthening the Capacity of Green Financial
       Instruments in Leveraging Climate Capital ⋯⋯⋯⋯⋯⋯⋯⋯⋯⋯ 97
   (Ⅳ) Gradual Internationalization of China's Green Investment
       Practice ⋯⋯⋯⋯⋯⋯⋯⋯⋯⋯⋯⋯⋯⋯⋯⋯⋯⋯⋯⋯⋯⋯⋯⋯⋯ 100

III. **Climate Financing Process and Cooperation of Other**
    Developing Countries ⋯⋯⋯⋯⋯⋯⋯⋯⋯⋯⋯⋯⋯⋯⋯⋯⋯⋯ 112
   (Ⅰ) Progress of Climate Financing in Other Developing Countries ⋯ 112
   (Ⅱ) Developing Countries Actively Conducting Climate Aid and

Cooperation ·· 119
(Ⅲ) Demands of Developing Countries and China's
    Countermeasures ·· 121

IV. Trump Presidency's Impact on Global Climate Financing ·········· 124
(Ⅰ) Influence on Global Climate Capital Flow ······························ 125
(Ⅱ) Changes to Global Climate Governance Pattern ···················· 129
(Ⅲ) New Role of China in Global Climate Financing ···················· 133

V. Cooperation of Multilateral Development Banks ······················ 136
(Ⅰ) High Potential of Cooperation among Multilateral
    Development Banks ·· 136
(Ⅱ) Cooperation between AIIB and EIB ······································ 139

VI. Policy Suggestions ·· 147
(Ⅰ) Stick to the Paris Agreement and Promote the Establishment of
    Global Integrated Climate Public Goods Supply Mechanism ··· 147
(Ⅱ) Promote the Role and Innovation of Emerging Multilateral
    Institutions in Addressing Climate Change ···························· 148
(Ⅲ) Starting from South–South Cooperation on Climate Change to
    Strengthen China's Soft Power of Climate Financing ············ 150

I. Progress in Global Climate Financing

(I) Continuous Expansion of Global Climate Financing Gap

The *Paris Agreement* approved at the Paris Conference on Climate Change on 12 December 2015 proposed to keep the global temperature rise this century within 2℃ above pre-industry level, and to pursue efforts to limit the increase even further to 1.5℃. To achieve this goal, countries will need to invest heavily in climate financing. The global climate financing gap is continuously expanding. According to the forecast of the International Energy Agency (IEA), the energy sector alone will need USD 16.5 trillion in 2015 – 2030 in order to achieve the goal of the *Paris Agreement*. Furthermore, according to the forecasting methodology of Sam Fankhauser et al., the additional amount of money to meet mitigation and adaptation needs in the next decade may be as high as USD 630 billion, with the demand for China's climate finance reaching USD 205 billion per year.[1] According to the statistics of Intergovernmental Panel on Climate Change (IPCC), the adaptation needed by developing countries will be about USD 70 ~ 100 billion per year from 2010 to 2050. United Nations Environment Program (UNEP) estimates that the annual adaptation capital need will reach USD 140 ~300 billion by 2030 and USD 280 ~500 billion by 2050. In addition, according to the estimation from McKinsey, the climate finance gap of sustainable infrastructures from 2015 to 2030 will reach USD

---

[1] Sam Fankhauser, Aditi Sahni, Annie Savvas & John Ward. Where are the gaps in climate finance? [J]. 2016,8(3). Climate and Development, 2016.

39~51 trillion, with the gap of middle-income countries accounting for 65%[①]. Although the forecast basis and methods are different, the forecast results all show that the global climate finance gap continues to expand.

According to the capital arrangement proposed in the *Paris Agreement*, developed countries commit themselves to mobilizing at least USD 100 billion per year to developing countries by 2020. According to the report published by the Organization for Economic Cooperation and Development (OECD), developed countries have mobilized climate finance of USD 52.2 billion in 2013 and USD 61.8 billion in 2014 for developing countries, with the public capital accounting for USD 37.9 billion and USD 43.5 billion respectively[②]. This counting methodology has been criticized by many countries. At the Bonn Conference on Climate Change in November 2017, developed countries had still not reached an agreement on the capital arrangement of USD 100 billion. Even if the private capital and ODA are counted as part of developed nations' financial commitment to developing nations, and the amount of USD 100 billion mobilized by developed countries is available in 2020, such capital is far from enough to cope with the great challenges brought by climate change.

(II) Integrated Capital Supply Mechanism on Global Climate Public Goods Not Formed

In the past 25 years, the supply channels of the global climate capital have been expanding, and the climate financing system is increasingly

---

① McKinsey. 2016. Financing Change: How to mobilize private sector financing for sustainable infrastructure. *The difference of USD 39 trillion and USD 51 trillion is calculated with different methods. USD 39 trillion is a conservative estimation, removing Chinese high growth of 13.4% in infrastructure investment and calculated based on the global average growth rate of 1.8%; USD 51 trillion is a relatively radical estimation, assuming China maintains its historical growth rate, which will raise the global average growth rate by 4.3%.*

② OECD, CPI. Climate Finance? in 2013-14 and the USD 100 billion goal [EB/OL]. http://www.oecd-ilibrary.org/docserver/download/9715381e.pdf?expires=1517901239&id=id&accname=guest&checksum=75A4697E7FA9ADF080FE02BD7169CB4B

evolving in a complex direction. Early climate financing depended heavily on the Official Development Aid (ODA) system and emphasized the one-way aid model of "donor countries to recipient countries". After the financial crisis, the growth of ODA in developed countries is slow, and the influence of emerging market countries and private financing mechanisms is gradually increasing. Although the multilateral climate mechanism based on the OECD aid system can benefit the world, its contribution to climate capital is the smallest in all channels. The bilateral channel is the main source of climate capital. Because it is efficient and direct, the bilateral channel is welcomed by the recipient countries, and its contribution to the climate capital is on the rise. In recent years, GCF has been established as an exclusive financing institution of UNFCCC. In addition, the establishment of the Africa Risk Capability (ARC) and the Caribbean Catastrophe Risk Insurance Facility (CCRIF) signs that the financing measures and risk management mechanism in the field of adaptation has made breakthrough progress (the architecture is shown in Figure 1.1). However, there are still many problems in the current global climate finance system, such as fragmentation and low coordination, and it is difficult to reflect the needs and contributions of emerging market countries and other developing countries, also it is difficult to support the sustainability and fairness of global public goods supply.

The international public climate funds are mainly derived from the ODA that can bring mitigation and adaptation effects. The proportion of bilateral climate funds to ODA has risen from 4% in 2003 to 18% in 2013~2014 years[1]. Rich countries transfer financial funds to poor countries via ODA in the traditional international aid system with the intention to establish a one-way relationship of "Donor countries – Recipient countries", meanwhile,

---

[1] Tracy C, Jan K, Annaka P. Climate finance shadow report [R/OL]. [2017-06-10]. https://www.oxfam.org/sites/www.oxfam.org/files/file_attachments/bp-climate-finance-shadow-report-031116-en.pdf.

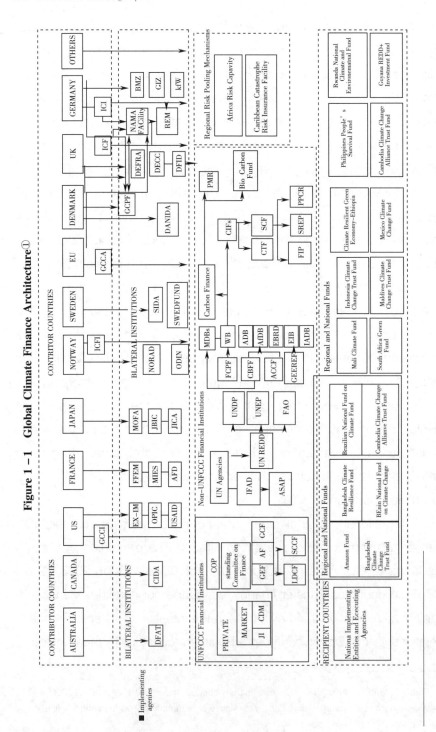

Figure 1-1 Global Climate Finance Architecture①

① WRI. The Future of the Funds: Exploring the Architecture of Multilateral Climate Finance [R/OL]. https://www.wri.org/sites/default/files/The_Future_of_the_Funds_0.pdf

I. Progress in Global Climate Financing

the former also has strong influence in the political and social fields of the latter. Therefore, ODA is essentially the extension of public affairs of the financial sector of the donor countries in international affairs. However, the climate financing mechanism should be one of the global public finance mechanisms that support the supply of global public goods and the goal is to maximize the benefits of funding in support of global mitigation and adaptation and to achieve the efficiency and fairness of the capital utilization globally①. Thus, there are differences between the two in the starting point and the focus of intervention. Even if there is a certain degree of synergy in some areas of development aid and climate financing, it is potentially possible to address climate change at the expense of development goals when taking ODA as the main part of climate capital. On the other hand, as the impact of regional and global risks deepens, the realization of the development goals of a single country is increasingly dependent on the supply of global public goods, especially in the field of climate change, but there is still lack of a more globally inclusive and integrated supply mechanism for public goods②.

In the climate finance system, which is highly dependent on ODA, the scale of public funds from developed countries is relatively fixed. After the financial crisis, in order to achieve the established performance goals, the international financial institutions are more inclined to mitigation projects with high efficiency, large scale and relatively high return on investment. Over the past 15 years, the global climate financing institutions have invested USD 11.7 billion in climate change field, supporting 697 projects, and mobilizing private capital and capital through other channels of USD 70

---

① Delina L. Multilateral development banking in a fragmented climate system: shifting priorities in energy finance at the Asian Development Bank [J]. International environmental agreements: politics, law and economics, 2017, 17(1): 73–88.

② HU Angang, ZHANG Junyi, GAO Yuning. Foreign assistance and national soft power: China's status quo and countermeasures[J]. Wuhan University journal (arts & humanity edition), 2017 (3): 5–13.

billion. But most of the funds are still inclined to invest in large projects rather than small projects in the most vulnerable areas or needed by the most vulnerable groups. In the existing projects of GCF, more than 90% of the expenditure is distributed through the United Nations Development Program (UNDP), the world bank, the Inter-American Development Bank and other institutions, rather than through the national or regional institutions that are in the most urgent need of support. The 3-year plan recently released by CIF shows that this trend is still strengthening. It is lack of mechanism design and performance goals for global public target investment in the most vulnerable areas and for the most vulnerable groups in the existing performance management system of these institutions. This is also an important reason why small projects in the areas of adaptation, energy efficiency and in the most vulnerable areas have always been difficult to obtain funds.

(III) Synergy of Different Finance Mechanisms Needs to be Improved

After years of development, there has been an increase in funding and tools of climate adaptation. However, even under the UNFCCC framework, the synergy issue still needs to be addressed. For example, Adaptation Fund (AF), Least Developed Country Fund (LDCF), and the Green Climate Fund (GCF) mainly support small-scale adaptation projects and their support scope overlaps to some extent. AF's project approval process is short and has accumulated years of experience in providing funds directly to institutions in developing countries. Nevertheless, the financing costs of AF are relatively high. Its funding sources are unstable and the amount of fund is small. AF can only support small-scale adaptation activities and it is difficult to ensure the sustainability of the project. LDCF and GCF are more suitable for focusing on large-scale projects with synergy effects and long-term support. However, GCF requires institutions to prove that projects have the ability of additional financing and has high requirements for risk control. There also exists concrete obstacles, for example, it is not adapted to the laws and management systems and lacks effective business patterns in the least developed regions. [1][2][3]

(Ⅳ) **The New and Old Multilateral Development Institutions Need to Seek Synergistic Mechanism and Model Innovation**

The role of the Multilateral Development Banks (MDBs) in the area of mitigation and adaptation to climate change is still important. According to the report jointly released by the six largest multilateral development banks in the world, MDBs made a joint climate capital commitment of USD 25 billion in 2015, including the internal capital of such MDBs and the external capital flowing into developing countries through the channels of MDBs[4]. Compared with the contribution of USD 25.7 billion in 2014, there was a slight downward trend[5], but the total investment in climate-related projects was higher than that in fossil fuels[6]. In addition, MDBs made a commitment in 2015 to increase the capital for climate-related projects by nearly USD 400 billion per year by 2020, however this figure only accounts

---

[1] LIU Jing. A new paradigm of international development cooperation reshaped in the plight of the global governance of aid[J]. Journal of international relations, 2017, 4:3.

[2] Strange A M, Dreher A, Fuchs A, et al. Tracking underreported financial flows: China's development finance and the aid-conflict nexus revisited[J]. Journal of conflict resolution, 2017, 61(5): 935-963.

[3] Delina L. Multilateral development banking in a fragmented climate system: shifting priorities in energy finance at the Asian Development Bank [J]. International environmental agreements: politics, law and economics, 2017, 17(1): 73-88.

[4] The statistical data is mainly of six largest MDBs, i. e., World Bank Group (WBG), African Development Bank (AFDB), Asian Development Bank (ASDB), Inter-American Development Bank (IADB), European Investment Bank (EIB) and European Bank for Reconstruction and Development (EBRD); Joint report on MDB, 2015.

[5] UNFCCC. Compilation and synthesis of the biennial submissions from developed country Parties on their strategies and approaches for scaling up climate finance from 2014 to 2020 [EB/OL]. http://unfccc.int/resource/docs/2016/sbi/eng/inf10.pdf

[6] E3G. Greening financial flows, what progress has been made in the development banks? [R/OL]. https://www.e3g.org/library/greening-financial-flows-what-progress-has-been-made-development

for less than 20% of the total investment made by MDBs①. In fact, the gap between the global public capital flow and the fund demand for realization of sustainable development targets will never be bridged by public funds, which is also pushing the method of the performance assessment in multilateral development institutions to make a change. In the past multilateral institutions mainly measured the performance by the amount of capital provided or paid, however, recently more and more institutions take the degree that public funds are used to leverage the resources of the private sector and donated domestic internal resources to measure the performance. The existing multilateral development banks are global asset, and they can and should be able to play a greater role in facing global challenges through mutual collaboration and further cooperation with emerging institutions.

In recent years, new development aid channels have continuously emerged, and the scale of aid from charity funds and non – governmental organizations has increased. The private commercialization capital flow in the climate area has substantially surpassed government funds. Emerging market countries can provide underdeveloped neighboring countries with official finance and technology support. Emerging multilateral organizations and regional development banks have exceeded the previous multilateral assistance framework led by traditional multilateral development banks in terms of aid funds. For example, China's current financial amount of assistance to Africa has exceeded that of the World Bank. From the available statistics, there are 14 funds invested and established by China or with foreign countries, with the amount of nearly USD 140 billion. These funds are mainly invested in developing countries, including Africa area, Latin America, countries along the Belt and Road, and ASEAN countries. In addition to the two green theme funds, South – South Cooperation Fund

---

① WRI. The Future of the Funds: Exploring the Architecture of Multilateral Climate Finance[R/OL]. https://www.wri.org/sites/default/files/The_Future_of_the_Funds_0.pdf

on Climate Change and the U. S. – China Green Fund, other non – green theme funds have also started to implement the concept of green investment. Emerging market countries have also accumulated a large number of local cases and experience in countries and regions that are fragile to climate, and have gradually emerged advantages in climate aid and investment.

Meanwhile, traditional multilateral institutions have been criticized that the design of their systems and mechanisms requires innovation and improvement. First of all, the current climate investment is too conservative. Overall, the loan – to – capital ratio of traditional multilateral institutions basically maintains below 4:1, which greatly limits the ability of their capital supply. In general, the loan – to – capital ratio of commercial banks is about 10:1. In fact, 7:1 is also considered to be a more appropriate proportion internationally. According to the study of ODI, if the multilateral development banks increase this ratio to 5:1, it will be able to release additional USD 200 billion of capital, and if it increases the ratio to 7.5:1, it can release USD 380 billion, which can inject blood into the globally extremely scarce climate fund system. Second, the complexity of the process rules when the multilateral funds approve projects is also one of the major obstacles currently impeding its role in climate finance. Different funds have different rules for examining and approving projects. The countries that apply for funds must spend a lot of time and effort in learning the rules and regulations of each fund, which greatly reduces the efficiency. Therefore, it is extremely necessary for the multilateral climate fund to adopt uniform application requirements and safeguard measures. However, multilateral financial institutions have strong research capabilities and wide vision of industry policies. They can provide global cases of success and failure, and can promote multi – level and multi – stakeholder communications to clarify the advantages and roles of public and private funds and multi – level agencies, which is precious for developing countries.

After the *Paris Agreement*, the new and old multilateral institutions are faced with a new pattern of competition and cooperation. On one hand, most developing countries are experiencing rapid development and still need to be supported by the financing channels of international capital markets through multilateral climate financing mechanisms. On the other hand, both developed countries and emerging market countries need to strengthen global and regional risk management capabilities through multilateral institutions to prevent accumulated national development achievements from being jeopardizing by the lack of global supply for public goods. For instance, regional institutions such as the AIIB, South – South funds, and silk funds need to focus on formulating regional agreements for public goods and services, such as support for the ASEAN Economic Community, regional infrastructure corridors, and Asian carbon market plans. The climate investment around emerging markets and aid countries or regions needs more integration with long – term accumulated local experience such as assistance from emerging market countries and foreign investment[①]. Therefore, how to promote the advantages of new and old multilateral institutions in the new structure and collaborative innovation is also an important issue in the development of the climate financing system.

Case Study I

Caribbean Catastrophe Risk Insurance Facility

In late September 2016, a Level –5 hurricane Matthew landed in the eastern Caribbean region, affecting countries such as Saint Lucia, Jamaica

---

① Dorsch M J, Flachsland C. A polycentric approach to global climate governance [J]. Global environmental politics, 2017, 17 (2) :45 –64.

I. Progress in Global Climate Financing

and Haiti and causing heavy rainfall. In Saint Lucia, most parts suffered from floods, some area even experiences landslides, and as much as 85% of farms suffered damage and loss. Saint Lucia is one of the participating countries in the Caribbean Catastrophe Risk Insurance Facility (CCRIF). According to calculation, the Rainfall Index Loss in this region was above the Attachment Point of the country's Excess Rainfall policy, and CCRIF made payout totaling USD 3,781,788 to Saint Lucia within 14 days after the occurrence of the hurricane. In addition, 31 residents in Saint Lucia had purchased the Livelihood Protection Policy under the Caribbean Climate Risk adaptation and Insurance projects, and therefore received a total payout of USD 102,000 after the hurricane. The Livelihood Protection Policy is a microinsurance product against extreme weather risks specially designed for and provided to low-income population in CCRIF.

Established in 2007, CCRIF is an insurance instrument providing the Caribbean governments (Central American governments have been included since 2015) with risk management services against natural disasters such as extreme weathers and earthquake. CCRIF has three projects, i.e., Integrated Sovereign Risk Management, Climate Risk Adaptation and Insurance, and Economics of Climate Adaptation. Aiming at providing the governments in the Caribbean affected by hurricane and earthquake with rapid and timely short-term liquidity, CCRIF is currently the only regional fund using parameter insurance in the world, enabling the governments in the Caribbean to buy hurricane and earthquake insurance at the lowest price. The establishment of CCRIF represents a change in the governments' approach to dealing with the risks of natural disasters.

The initial capital of CCRIF comes from the donation of Japanese Government, and the follow-up capital comes from the Multi-donor Trust Fund contributed by the governments of Canada, the European

Union, Britain, France, Ireland and Bermuda, the World Bank and the Caribbean Development Bank, as well as the membership fees paid by participating governments. In order to guarantee the repayment capacity of CCRIF, CCRIF puts the risks of each country into a risk pool and models scenarios to make ensure that the fund could pay the compensation for one in 10,000 chance events in a given year. Basically, the occurrence of the events with a one in 10,000 chance means that some major economies in the Caribbean will suffer greatly at the same time. For example, Jamaica, Barbados, Trinidad, Cayman and Bahamas are all hit by natural disasters in a single given year. Though such events are unlikely to happen, CCRIF includes such possibilities in its financial modeling and has the capacity of meeting claims from participants if such an event should occur.

Case Study II

Africa Risk Capacity

It is estimated that the direct economic losses resulting from the possible drought disaster in Sub–Saharan Africa may be as high as USD 3 billion, which is to a huge burden for both the aiding government and the local governments. ①

Malawi suffered from severe drought in 2015–2016 and the country was in a state of disaster because of food shortage. Luckily, Malawi is a participant of Extreme Climate Facility from Africa Risk Capacity (ARC), which made payments totaling USD 8.1 million to Malawi based on the impact of the drought and the coverage of the insurance, thus making up for some of the financial losses. In this case, though the availability time

---

① ARC official website: http://www.africanriskcapacity.org/2016/10/29/vision–and–mission/

I. Progress in Global Climate Financing

of the insurance capital was delayed due to technical factors, it reduced the fiscal burden of the government significantly as a whole.

ARC is a finance mechanism established in 2012 by African Union, aiming at helping member countries increase their capacity in planning, preparing and responding to extreme weather accidents and other natural disasters. Currently, ARC has three products: Extreme Climate Facility (ECF), Outbreak & Epidemic (OE) and Replica Coverage (RC).

ARC uses Africa Risk View, an advanced satellite weather surveillance software developed by World Food Program, to quantitatively assess the situations of African countries affected by bad weather and make payouts according to the corresponding mechanisms. The initial capital comes from the premiums of participating countries and the contributions of donors. Countries can choose the corresponding rate from ARC according to coverage they need. ARC provides reinsurance for this venture capital, and also uses insurance capital for investment activities. When a disaster occurs, ARC compensates member countries according to the predetermined and transparent payout rules. Because disasters usually will not break out in all parts of Africa at the same time in a single year, the establishment of disaster risk pools such as ARC may not only greatly reduce the emergency cost of Africa, but also substantially reduce the reliance on foreign aid.

II. Progress of Climate Financing in China

In recent years, with the constant promotion of climate change policies, the wide adoption of public – private partnership (PPP) mode, and the gradual implementation of green finance, China's climate finance channels keep enriching. Meanwhile, by setting up green funds through cooperation with other developing countries, China's green investment practice begins to go abroad, not only providing climate finance to the world, but also enriching the domestic capital sources.

(I) Gradual Improvement of Climate Change Policy System

Since 2007 when the State Council issued *China's National Climate Change Programme*, the Chinese Government has successively adopted a series of measures to actively respond to climate change, and gradually establish the climate change policy systems covering both mitigation and adaptation, which provides powerful policy support for the domestic climate finance supply.

1. Carbon emission intensity control

The carbon emission intensity per unit GDP was included as a binding indicator in the development plan for the first time in the 12th Five – Year Plan, indicating that low – carbon transformation has become one of the important goals of China's social and economic development. The State Council issued *Work Plan for Controlling Greenhouse Gas Emissions During the 12th Five – Year Plan Period* to reduce emissions per unit GDP to 31 provincial governments, and a series of documents to promote the

II. Progress of Climate Financing in China

normalization and standardization of greenhouse gas statistics, so as to assess and evaluate in a timely manner the completion of carbon emission reduction per unit GDP. Under the guideline of national policies, provincial governments have further sub-assigned the responsibility of emission control to municipal governments. During the 13th Five-Year Plan Period, the control over carbon emission reduction per unit GDP is continuously included in the goal system of the *Outline of the 13th Five-Year Plan*, indicating that China has gradually formed a multi-level carbon emission intensity control system based on administrative regions.

Figure 2 – 1   Overview of Carbon Emission Intensity Control Policies

October 2010	Outing of the 12th Five-Year Plan
May 2011	Guidance for Compiling Procincial Greenhouse Gas Emission Lists (Trial)
December 2011	Work Plan for Controlling Greenhouse Gas Emissions During the 12th Five-Year Plan Period
May 2014	Opinions on Improving Response to Climate Change and Statistical Work for Greenhouse Gas Emissions
August 2014	Measures for Performance Evaluation and Accountability for the Target of Lowering Carbon Dioxide Emissions per Unit of GDP
January 2015	Nation on Launching Procincial GHG Inventory Compilation in the Next Phase
March 2016	Outline of the 13th Five-Year plan
November 2016	Work Plan for Controlling Greenhouse Gas Emissions During the 13th Five-Year Plan Period

2. Carbon emissions trading market

In November 2011, National Development and Reform Commission (NDRC) approved seven provinces and cities (Beijing, Shanghai, Tianjin, Chongqing, Guangdong, Hubei and Shenzhen) to launch the carbon emissions trading pilots. Since 2013, each carbon trading pilot has been launched, and now the carbon trading system, which accords with the actual situations of the region, has been preliminarily formed. By 2016 the total cumulative transaction amount of the seven pilot carbon trading markets is 116 million tons with RMB 2.5 billion, and the transaction amount in 2016 alone is RMB 1 billion, 22.1% higher than that of 2015[①]. According to the public information of each trading exchange, the accumulative trading volume of Chinese Certified Emission Reductions (CCER) in 2016 was about 53 million tons of carbon dioxide equivalent and there were a total of 861 registered CCER projects. Compared with 2015, there were 409 newly registered projects in 2016. As of the end of 2017, the total cumulative transaction amount of the seven pilot carbon trading markets is 182 million tons with RMB 3.6 billion[②], and the transaction amount in 2017 alone is about RMB 1.1 billion, 10% higher than that of 2016. These indicate a rapid development trend of Chinese carbon markets.

In 2014, NDRC launched the system design research on national carbon market. The *Interim Measures on the Management of Carbon Emissions Trading* issued in December 2014 proposed many normative requirements on the development directions, organizational structure system, etc. of national carbon markets, and the *Notice on Implementing Key Work on Launching National Carbon Emissions Trading Market* issued in 2016 further specified the covered industries and specific launching requirements of national carbon markets. On December 19th 2017 China launched the national carbon emissions market, which is helpful for the full release of the climate finance supply capacity of carbon market.

---

① Beijing Environmental Exchange, Beijing Green Finance Association. (2017). 2016 Beijing Carbon Market Annual Report. http://www.cbeex.com.cn/article//xxfw/xz/bjtscndhq/201701/20170100059897.shtml

② Beijing Environmental Exchange, Beijing Green Finance Association. (2018). 2017 Beijing Carbon Market Annual Report. http://files.cbex.com.cn/cbeex/201802/20180211162427630.pdf

II. Progress of Climate Financing in China

Table 2-1 Overview of China's Carbon Trading Related Policies

Date of Issuance	Relevant Policies and Actions	Relevant Contents
October 2011	Notice on the Pilot Work on Carbon Emissions Trading	Approving seven provinces and cities, i. e. Beijing, Shanghai, Tianjin, Chongqing, Guangdong, Hubei and Shenzhen, to launch the carbon emissions trading pilot work in 2013
June 2012	Interim Regulation of Voluntary Greenhouse Gas Emission Trading	Performing systematic specification on the development, trading and management of CCER projects
July 2012	Implementation Plan for Assessing the Responsibilities on Energy Conservation Goals of Ten Thousand Enterprises	Starting the construction of energy saving quantity trading markets, with Shandong, Jiangsu and Fujian having already established provincial energy saving quantity trading markets
October 2012	Guidelines for Validation and Verification of CCER Projects	Specifying the filing requirements for validation and verification institutions of CCER projects
November 2013	National Strategy for Climate Change Adaptation	Dealing with global climate change and coordinating the national climate change adaptation work
December 2014	Interim Measures on the Management of Carbon Emissions Trading	Proposing normative requirements on the development directions, organizational structure design etc. of national unified carbon emissions trading market
September 2015	Integrated Reform Plan for Promoting Ecological Progress	Deepening the construction of carbon emissions trading pilots, and gradually establishing national carbon emissions trading market
September 2015	China – U. S. Joint Presidential Statement on Climate Change	Planning to launch the national carbon emissions trading market in 2017, covering the major industries including iron and steel, power, chemistry, building materials, paper – making and nonferrous metals etc
January 2016	Notice on Implementing Key Work on Launching National Carbon Emissions Trading Market	Specifying eight industries participating in national carbon market, requiring the verification on the historical carbon emissions of incorporated enterprises, and proposing the verification report on the supplementary data of carbon emissions of the enterprises

continued

Date of Issuance	Relevant Policies and Actions	Relevant Contents
January 2016	Notice on the Trial Implementation of Issuance and Voluntary Purchase of Green Power Certificate of Renewable Energy	To be started when suitable from 2018
August 2016	Exposure Draft of Management Measures for Carbon Allowance of New Energy Vehicles	Proposed operation time to be determined
November 2016	Work Plan for Controlling Greenhouse Gas Emissions During the 13th Five – Year Plan Period	Performing comprehensive deployment on climate change work and promoting low carbon development during the 13th Five – Year Plan period
December 2016	Notice on Release of Green Development Indicator System and Assessment Target System of Ecological Civilization Construction	Taking carbon emissions reduction as the basis for evaluation and assessment of ecological civilization construction
December 2016	Development Plan of Energy Conservation and Environmental Protection Industries During the 13th Five – Year Plan Period	Developing energy conservation and environmental protection industries and strengthening pollution prevention work
January 2017	First Biennial Report on Climate Change of the People's Republic of China	1) Comprehensively renewing the national greenhouse gas list 2) Systematically summarizing and analyzing the mitigation actions and effects of China during the 12th Five – Year Plan Period 3) Renewing the capital, technology and capacity building requirements as well as the subsidies obtained 4) Reporting the domestic measurement, report and verification (MRV) system for the first time

continued

Date of Issuance	Relevant Policies and Actions	Relevant Contents
August 2017	Pilot Program for Paid Use and Trading System of Energy Consumption Rights	Specifying the piloting in Zhejiang, Fujian, Henan and Sichuan since 2017, and gradually promoting it since 2020. Currently, the four pilot provinces are at the system design stage
December 2017	National Carbon Emissions Trading Market Construction Program (Power Generation Industry)	Commit to complete the infrastructure construction within one year, to complete the simulation trading within one year, and to slowly expand market coverage, products and means of transaction based on the stable operation of carbon market of power sector

3. Multi – level Low – Carbon Pilots

During the 12th Five – Year Plan period, China actively carried out various low – carbon pilots, and had initially formed a multi – level low – carbon pilot pattern covering provinces/autonomous regions, cities, industrial parks, towns and communities.

Since 2010, NDRC has begun to promote the pilot work of low – carbon provinces, autonomous regions and cities. By 2017, China has carried out three batches of low – carbon province/autonomous region pilots, covering 29 provinces/autonomous regions and 81 cities, with the pilot work carried out across the country. Since 2012 when the second batch of low – carbon province/autonomous region pilots is announced, China has begun to require all pilots to set corresponding emission peak goals to form a forcing mechanism. As of October 2017, 34 provinces, autonomous regions and municipalities have proposed the annual goal of achieving carbon emission peaks, with 12 cities such as Beijing, Shanghai, Guangzhou and Hangzhou putting forward clearly the plan of achieving carbon emission peaks before 2020. According to the relevant review, the carbon emission reduction per unit GDP in the low – carbon pilots during the 12th Five – Year Plan period, indicating that low-carbon pilot has received initial success. ①

In order to explore the new low-carbon development mode of industry, in September 2013, MIIT and NDRC jointly issued *Notice on Carrying out National Low-Carbon Industrial Park Pilots*, since then China's low-carbon industrial park pilot work has launched officially. Currently, 51 low-carbon industrial parks have been approved. During the pilot period, the energy consumption and carbon emission per unit industrial added value decreased dramatically, and the carbon productivity was effectively improved. ②

In August 2015, the NDRC issued *Notice on Accelerating National Low-Carbon City (Town) Pilots*, determining eight towns as the first batch of national low-carbon city pilots, aiming at exploring the low-carbon development modes in the new urbanization process throughout the entire period from planning to construction, operation and management.

In March 2014, NDRC issued *Notice on Carrying out Low-Carbon Community Pilots*, starting the pilot work of low-carbon communities. In February 2015, NDRC issued *Guidelines for Low-Carbon Community Pilot Construction*, further guiding the low-carbon pilot work of different types of communities such as existing communities and rural communities, and promoting the exploration of community-based carbon emission control paths. At present, there are about 800 low-carbon community pilots across the country. ③

4. Adaptation City Pilots

The Chinese Government is increasingly concerned about climate change. After the concept of adaptation to climate change was firstly

---

① Brookings-Tsinghua Center for Public Policy. China's Low-Carbon Development Report 2017[R].

② Institute of Urban Development and Environmental Studies, Chinese Academy of Social Sciences. Construction Practice and Innovation of National Low-Carbon Industrial Parks[R].

③ Brookings-Tsinghua Center for Public Policy. China's Low-Carbon Development Report 2017[R].

II. Progress of Climate Financing in China

proposed in *China's Agenda* 21 issued in 1994, it was clearly specified in the 12th *Five – Year Plan* in 2010 to "take the factor of climate change into full consideration during productivity deployment, infrastructure and key project planning design and construction, and increase the level of adaptation to climate change in key fields such as agriculture, forestry, water resources and coastal and ecologically vulnerable areas." In 2013, NDRC, Ministry of Finance, Ministry of Housing and Urban – Rural Development, Ministry of Transport, Ministry of Water Resources, Ministry of Agriculture, State Forestry Administration, China Meteorological Administration and State Oceanic Administration jointly issued *National Strategy for Climate Change Adaptation*.

In order to actively explore the urban solutions for adapting to climate change, in January 2016, NDRC and Ministry of Housing and Urban – Rural Development jointly developed *Action Plan for Urban Adaptation to Climate Change*, proposing a series of development goals and tasks from the perspectives of urban planning, infrastructure, architecture, ecological afforestation, water and risk management. Different regions have started to carry out active explorations. In February 2017, authorities jointly issued the *Notice on Issuing the Pilot Work on Climate Adaption City Construction*, specifying the work goal of including climate change adaptation concepts into urban planning and construction management process, determining 28 regions such as Hohhot in Inner Mongolia, Dalian in Liaoning and Wuhan in Hubei as the first batch of climate adaptation pilot cities and the tasks of the pilot cities are clarified from the aspects of urban adaptation concepts, climate change and climate disaster monitoring and early warning capacity, key adaptation actions, policy experimental base and international cooperation platforms. Since then, the pilot work of China's adaptation cities has been officially launched.

Figure 2-2 Relevant Policies and Actions of Low-Carbon Pilots

	August 2010	Notice on Launching Low-Carbon Pilot Procinces and Cities
Notice to Carry Out Second Batch of Low-Carbon Provinces City Pilot Work	December 2012	
	Septmber 2013	Notice on Carrying out National Low-Carbon Indusrrial Park Pilots
National Strategy for Climate Change Adaptation	November 2013	
	March 2014	Notice on Carrying out Low-Carbon Community Pilots
Guidelines for Low-Carbon Community Pilot Construction	February 2015	
	August 2015	Notice on Accelerating National Low-Carbon City (Town) Pilots
Action Plan for Urban Adaptation to Climate Change	January 2016	
	January 2017	Notice to Carry Out Third Batch of Low-Carbon City Poilt Work
Notice on Issuing the Pilot Work on Climate Afaption City Construction	February 2017	

(II) Further Deepening the Adoption of PPP Mode in Green and Low-Carbon Fields

In recent years, the policy support to public-private partnership (PPP) has been constantly strengthened, particularly in green and low-carbon fields. The *Notice on Further Promoting Public-Private Partnership Work in Public Services* issued by the Ministry of Finance in October 2016 pointed out that in the public service fields such as urban rail transit, clean energy facility, garbage disposal and sewage treatment, due to high degree of marketization and mature adoption of PPP mode, new projects in different regions should "be forced" to adopt the PPP mode.

According to the Ministry of Finance's PPP project information database, as of the end of December 2017, the total PPP project number is 7,137 with investment amount of RMB 10.8 trillion, implying a net increase

II. Progress of Climate Financing in China

of 359 projects and RMB 637.6 billion in the quarter and a year-on-year net increase of 2,864 projects and RMB 4.0 trillion, in which there are 2,729 projects at implementation stage with investment amount of RMB 4.6 trillion and an implementation rate of 38.2% (means the ratio of the number of projects that have launched to the total number of projects in the database). There are 3,979 projects in the field of pollution prevention & control and green & low-carbon, with investment amount of RMB 4.1 trillion, accounting for 55.8% and 38.0% of the total number in the database respectively, implying a year-on-year net increase of 1,507 projects and RMB 1.4 trillion.

It is important to give full play to the guiding role of PPP demonstration projects. Among the first and the second batches of PPP demonstration projects there are 105 green and low-carbon projects with investment amount of RMB 504.956 billion, accounting for 62.9% of the total investment. Most of such projects are green municipal projects. In 2016, the third batch of demonstration projects totaled 513, accounting for 44% of all declared projects with investment amount of over RMB 1.19 trillion. Compared with the second batch, the invested amount in ecological construction and environmental protection projects in the third batch increased significantly, with new investment of RMB 53.94 billion and 33 newly added projects, among which the comprehensive treatment projects have increased significantly. There are a total of 46 ecological construction and environmental protection projects in the third batch, with investment amount of RMB 81.056 billion, the number of projects accounting for 8.9%, and the investment amount accounting for 6.9%.

With the wide application of green PPP projects, the role of financial capital gradually shifts from direct investment to the leveraging of social capital investment, thus constantly strengthening the leverage effect. In 2016, the investment of national public finance in energy conservation and environmental protection was RMB 473.482 billion, with the capital from central finance continued to decrease while capital support of local finance showed a slight increase trend compared with 2015, accounting for 2.76% of the total expenditure of local finance. In terms of specific investment fields, national public finance used for pollution prevention and ecological protection was RMB 144.755 billion and RMB 32.654 billion respectively, increasing by 10.2% and 6.9% year on year respectively; Other expenditure on energy conservation and environmental protection was RMB 78.749 billion, an increase of 53.6% compared with that in 2015.

In 2016, the general public budget expenditure on national renewable energy was RMB 8.612 billion, decreasing by 47.7% year on year, which is mainly due to the decrease of budget on local finance, while the public budget expenditure of the central finance in this field increased by 418.1% year on year. In addition, the expenditure of national governmental funds on additional power price income arrangements for renewable energy was RMB 59.506 billion, increasing by 2.7% year on year.

Table 2-2 Comparison of the Expenditure of Central Finance and Local Finance on Climate Change and Other Fields

unit: RMB 100 million

Years	Central			Local		
	2014	2015	2016	2014	2015	2016
Energy conservation and environmental protection	2033.03	400.41	295.49	3470.90	4402.48	4439.33
Education	4101.59	1358.17	1447.72	21788.09	24913.71	26625.06
Science and technology	2541.81	2478.39	2686.10	2877.79	3384.18	3877.86
Culture, sports and media	508.47	271.99	247.95	2468.48	2804.65	2915.13
Medical service	2931.26	84.51	91.16	10086.56	11868.67	13067.61
Total expenditure	74161.11	80639.66	86804.55	151785.56	175877.77	160351.36
Proportion of energy conservation and environmental protection(%)	2.74	0.5	0.34	2.29	2.50	2.76

Source: Fiscal data of Budget Division, Ministry of Finance http://yss.mof.gov.cn/2016js/

II. Progress of Climate Financing in China

Table 2-3 Statistical Data of PPP Demonstration Projects in Green and Low-Carbon Related Industries

Field	First batch of demonstration projects		Second batch of demonstration projects		Third batch of demonstration projects		Total		
	Number of projects	Investment amount (RMB 100 million)	Number of projects	Investment amount (RMB 100 million)	Number of projects	Investment amount (RMB 100 million)	Number of projects	Proportion (%)	Investment amount (RMB 100 million)	Proportion (%)
1. Transportation:	9	1565.96	38	3491.78	40	4467.48	87	11.48	9525.22	47.98
Rail traffic	7	1526.51	13	2452.67			20	2.64	3979.18	20.05
Expressway			7	613.94	26	3689.36	33	4.35	4303.30	21.68
Non-toll road			8	232.38			8	1.06	232.38	1.17
Transportation hub			4	17.32	2	20.57	6	0.79	37.89	0.19
Airport			2	81.98	1	203.17	3	0.40	285.15	1.44
Railway			2	45.7	3	126.41	5	0.66	172.11	0.87
Public transportation			1	15			1	0.13	15.00	0.08
Bridge			1	32.78	4	370.09	5	0.66	402.87	2.03
Others	2	39.45			4	57.87	6	0.79	97.32	0.49
2. Municipal projects	8	77.27	66	810.53	119	1495.35	193	25.46	2383.15	12.01
Waste treatment	1	5.26	22	97.24	31	124.44	54	7.12	226.94	1.14
Underground utility tunnel	1	13	14	407.77	31	838.68	46	6.07	1259.45	6.34

95

continued

Field	First batch of demonstration projects		Second batch of demonstration projects		Third batch of demonstration projects		Total			
	Number of projects	Investment amount (RMB 100 million)	Number of projects	Investment amount (RMB 100 million)	Number of projects	Investment amount (RMB 100 million)	Number of projects	Proportion (%)	Investment amount (RMB 100 million)	Proportion (%)
Park			1	25	4	43.94	5	0.66	68.94	0.35
Gas supply			2	3.64	2	3.40	4	0.53	7.04	0.04
Heat supply	3	26.4	6	59.52	13	62.24	22	2.90	148.16	0.75
Water supply	3	32.61	18	190.11	24	152.65	45	5.94	375.37	1.89
Sponge city			1	13.85	5	208.90	6	0.79	222.75	1.12
Greening			1	5.5	4	16.22	5	0.66	21.72	0.11
Water drainage			1	7.9	5	44.88	6	0.79	52.78	0.27
3. Environmental protection:	11	111.99	31	742.46	82	932.79	124	16.36	1787.24	9.00
Wastewater treatment	9	59.08	15	337.49	40	181.37	64	8.44	577.94	2.91
Comprehensive environmental treatment	2	52.91	15	364.97	38	711.32	55	7.26	1129.20	5.69
Wetland protection			1	40	4	40.10	5	0.66	80.10	0.40
Total	28	1755.22	135	5044.77	241	6895.62	404	53.30	13695.61	68.99

Source: PPP Project Base of PPP Center, Ministry of Finance

II. Progress of Climate Financing in China

(III) Constantly Strengthening the Capacity of Green Financial Instruments in Leveraging Climate Capital

In 2016 and 2017, the construction of China's green financial system has been continuously promoted, the market size of various green financial instruments has been further expanded, and the financing channels for addressing climate change have been widened.

1. Green credit

As one of the earliest green finance products in China, green credit has played a fundamental role in China's green financial system. As of the end of 2016, the green credit balance of 21 major banks rose to RMB 7.51 trillion, increasing by 7.13% year on year and accounting for 8.83% of the total loan balance. The loan balance of energy conservation and environmental protection projects and services was RMB 5.81 trillion, and the loan balance for strategic emerging industries such as energy conservation and environmental protection, new energy and new energy vehicle was RMB 1.7 trillion. It is predicted that the annual conservation of standard coal will be 188 million tons and the reduction of carbon dioxide equivalence will be 427 million tons. In terms of investment fields, green transportation and clean energy projects are the major investment fields of China's green credit. [1]

China's green credit product innovation is ongoing and the financial leverage of green assets has been continuously strengthened. Products include the loan products with green rights and interests such as carbon assets and future earning right of Energy Performance Contracting projects,

---

[1] Source: Statistical data of CBRC

as well as the financing products directly supporting green projects. In addition, the development of green credit asset securitization will further release capital for green projects development.

2. Green securities

In 2017, the issuance of green bonds in China and abroad amounted to RMB 248.314 billion, accounting for 32% of the global circulation. Although the issuance amount is lower than that in 2016, China was still the world's largest green bond issuing market.[①] Figure 2-3 shows the purpose distribution of the capital raised by China's green bonds. As showed in the figure, except for financial bond, clean energy is the most

Figure 2-3 Purpose of the Capital Raised by China's Green Bonds

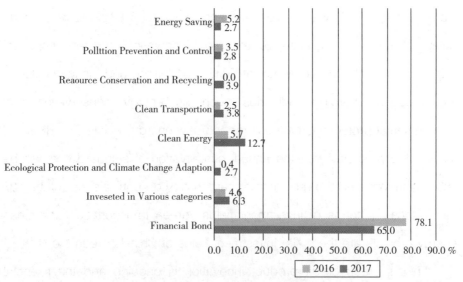

Source: International Institute of Green Finance, Central University of Finance and Economics (2018). 2017 *China Green Bond Market Report*.

---

① International Institute of Green Finance, Central University of Finance and Economics (2018). 2017 *China Green Bond Market Report*.

II. Progress of Climate Financing in China

important area for green bond raising in China in the last two years, mainly involving solar power, wind power and hydroelectric power. In addition, derivatives such as China Bond – China Climate – Aligned Bond Index and Central University of Finance and Economics (CUFE) – China Securities Green Bond Index are enriching day by day. In the near future, the development of trading – type open index fund products with green bond index as the subject is beneficial for mobilizing social capital to support the projects in the field of climate change.

3. Climate Catastrophe Insurance

The construction of climate catastrophe insurance system in China has made progress in recent years, and its function in enhancing the capacity of adapting to climate change and preventing poverty due to disasters has appeared gradually. Currently, several cities such as Ningbo, Shenzhen and Xiamen have launched a climate catastrophe insurance system. Taking Xiamen as an example, in the form of an insurance consortium formed by five insurance companies, all residents in Xiamen have life, housing and property insurance against natural disasters such as typhoons, rainstorms, floods and earthquakes.

The launching of fiscal catastrophe insurance products further strengthens the function of insurance in enhancing the regional climate resilience. As of November 2016, the Guangdong provincial fiscal catastrophe insurance has been fully implemented in ten pilot cities, providing an accumulative risk guarantee fund of RMB 2.347 billion; by August 2017, an accumulative amount of RMB 65.276 million has been paid in compensation. It plays an important role in the construction of natural disaster relief system.

4. Green Funds

As the first climate change policy fund in developing countries, the Clean Development Mechanism Fund (CDMF) of Ministry of Finance, has invested in 223 green and low – carbon projects through social fund management mode since the comprehensive operation began in 2010, realizing an emission reduction of 46.5462 million tons of carbon dioxide equivalence.① At local level, with vigorous promotion of national policies, green funds have become an important source of capital to promote urban green transformation, and the number of new funds has increased significantly. As of the end of 2016, there were 265 green funds registered at the China Funds Association, and 121 funds were established in 2016. Among all green funds filed, the number of green industry funds accounts for 83%, of which 51% are clean energy industry funds. This provides important sources for the emerging green industries that have not yet met the investment requirements of financing instruments such as green credit and green bonds.

(Ⅳ) Gradual Internationalization of China's Green Investment Practice

In recent years, China's concept of overseas green investment has been deepening. After the Ministry of Commerce and the Ministry of Environmental Protection jointly issued the *Guidelines on Overseas Investment in the Cooperation on Environmental Protection* in 2013, the Ministry of Environmental Protection issued the *Guidelines on Promoting the Construction of Green "Belt and Road"* in April 2017, aiming at implementing the concept of ecological civilization from the perspectives of infrastructure construction, trading and foreign investment, so as to promote the construction of the green silk road, and stress the importance of the international multilateral cooperation institutions and funds as green capital supplier initiated by China. In September 2017, seven institutions such as Green Finance Committee under China Society for Finance & Banking jointly issued the *Environmental Risk Management Initiative for China's Overseas Investment*, encouraging and guiding China's financial institutions to strengthen the environmental risk management in overseas

---

① Source: http://www.cdmfund.org/zh/ycsy/index.jhtml.

II. Progress of Climate Financing in China

investment and to follow the responsible investment principles.

Driven by policies and industrial initiatives, China is actively carrying out bilateral green cooperation with the majority of developing countries, aiming at realizing the win-win of participant countries on economic transformation development and environmental protection. The funds established by Chinese capital or Sino-foreign capital have become important platforms for green cooperation.

According to existing statistics, there are a total of 14 funds established by Chinese capital or Sino-foreign capital, totaling near USD 140 billion. Such funds mainly invest in developing countries, including African countries, Latin America countries, countries along the Belt and Road, and ASEAN countries. Except for two green concept funds, i.e. Climate Change South-South Cooperation Fund and Sino-US Green Fund, other non-green concept funds also begin to practice green investment concept, providing much valuable green investment experience.

Table 2-4 Overview of China-foreign cooperation fund

Name of the Fund	Date of Establishment	Scale of Fund Investment Commitment (Billions of dollars)	Focus Areas	Geographical range of investment
Asian				
Silk Road Fund	2014	54.5	Infrastructure construction, energy, production capacity	Countries along the One Belt and One Road, mainly Asian countries
China - ASEAN Investment Cooperation Fund	2009	10	Infrastructure construction, energy, natural resources	China, ASEAN
China - ASEAN Maritime Cooperation Fund	2011	0.5	Maritime economy, environmental conservation	China, ASEAN

continued

Name of the Fund	Date of Establishment	Scale of Fund Investment Commitment (Billions of dollars)	Focus Areas	Geographical range of investment
Euro–Asian Cooperation				
China – Central and Eastern Europe Investment Cooperation Fund	2012	11.5	Infrastructure construction, energy, manufacturing, communications	Central and Eastern Europe
Russia – China Investment Fund	2012	1(Russia) + 1(China) = 2	Infrastructure construction, agriculture, natural resources	70% invested in Russia, 30% in China
Latin America				
Sino–Latin American Production Capacity Cooperation Investment Fund	2014	20	Infrastructure construction, energy, natural resources, manufacturing, information and communication technology	Latin American countries
China–Latin America's infrastructure special fund	2014	10	Infrastructure construction	Latin American countries
China – Latin America Cooperation Fund	2014	5	Infrastructure construction, energy, natural resources, agriculture, manufacturing, information technology	Latin American countries
China – Mexico Investment Fund	2014	2.4	Infrastructure construction, automobile industry	Mexico
Africa				
China – Africa Development Fund	2007	3	Mining, energy, manufacturing	Africa
African Development Fund	2014	1(African Development Bank) +1 (China) =2	Infrastructure construction	Africa

II. Progress of Climate Financing in China

continued

Name of the Fund	Date of Establishment	Scale of Fund Investment Commitment (Billions of dollars)	Focus Areas	Geographical range of investment
China – Africa Industrial Capacity Cooperation Fund Company Limited	2015	10	Infrastructure construction, energy, manufacturing, agriculture, mining	Africa
Global				
South – South Climate Cooperation Fund	2015	3.2	Adaptation and mitigation of climate change	Developing countries
South – South Cooperation Assistance Fund	2015	2	No specific subject	Least developed countries, small countries, island countries
North America				
Sino – US Green Fund (Former name: China – US Building Energy Efficiency and Green Development Fund)	2016	3.05 (raised in the first batch)	Building energy efficiency, emission reduction, industrial structure upgrade	China (corporate with the municipal government)
Total number of funds: 14		Total scale of funding: USD 13.82 billion	The most concerned areas: infrastructure construction, energy cooperation	Main focus countries: developing countries

1. Green investment under the goal of regional economic integration: Silk Road Fund

Silk Road Fund was established in Beijing through the joint contribution of foreign exchange reserves, China Investment Corporation, China Development Bank and the Export – Import Bank of China on December 29, 2014, with a total capital of USD 40 billion. In 2017, China announced to increase capital by RMB 100 billion. As a medium – and long – term development and investment fund, Silk Road Fund aims to promote

103

projects in fields such as infrastructure, resource development, industrial cooperation and financial cooperation in countries along the Belt and Road through diversified financial instruments such as equity, bond and loan. As of March 2017, Silk Road Fund had 15 contracted projects, with an accumulative investment commitment of about USD 6 billion and involving a total investment of exceeding USD 80 billion.

By including social responsibilities such as environmental protection into project feasibility assessment and risk management system, Silk Road Fund has always carried out the idea of green investment while promoting interoperability. The current contracted green projects include the Karot Hydropower Project along China – Pakistan Economic Corridor, Yamal LNG Integrated Project and UAE Dubai Hassyan Clean Coal Power Station Project. Table 2 – 5 shows the overview of the milestones of green investment and financing of Silk Road Fund.

Table 2 – 5   Milestones of Green Investment and Financing of Silk Road Fund

2014	8 November	President Xi Jinping announced at the conference on strengthening interconnection partnership that China will contribute RMB 40 billion to set up the Silk Road Fund.
	29 December	Silk Road Fund was officially established in Beijing
2015	April	Silk Road Fund invested in Karot Hydropower Project, one of the projects with implementation priority along China – Pakistan Economic Corridor
	14 December	Silk Road Fund and Kazakhstan Ministry of Export Investment signed the Framework Agreement on the Establishment of China – Kazakhstan Productivity Cooperation Fund
	17 December	Silk Road Fund and Russian Novatek signed Yamal LNG Integrated Project trading agreement
2016	19 January	Silk Road Fund and Saudi International Power and Water Corporation signed a memorandum of understanding on joint development of UAE power stations.

II. Progress of Climate Financing in China

continued

2016	18 June	Silk Road Fund and Serbian Government signed a memorandum of understanding on new energy project cooperation.
	13 October	Silk Road Fund and IFC Asian Emerging Markets Fund Management Corporation signed a subscription agreement on IFC Asian emerging market fund projects.
2017	14 May	China will enhance the capital support to the construction of the Belt and Road, providing an additional capital of RMB 100 billion to Silk Road Fund.
	8 June	Chinese and Kazakhstan governments signed a duty–free agreement on the individual–type income from the direct investment of China – Kazakhstan Productivity Cooperation Funds in Kazakhstan.
	November	Signed a Cooperation Agreement on Joint Investment Platform for Energy Infrastructure in Beijing with GE Energy Financial Services, a subsidiary of GE, to jointly invest in infrastructure projects in power grid, new energy, oil and gas industries of countries and regions along the "One Belt and One Road" initiative.

2. Green Capacity Cooperation: Sino – Latin America Capacity Cooperation Fund

The Sino–Latin America Capacity Cooperation Fund was established in 2015 with the joint contribution of USD 30 billion by foreign exchange reserves and China Development Bank, with the capital for the initial period of USD 10 billion. The Fund aims to realize the complementary development for China and Latin America by promoting capacity cooperation between China and Latin America in fields such as manufacturing, high technology, agriculture, energy and mineral resources, and infrastructure.

In currently operating projects, there are good practices of capacity cooperation in the field of green and low–carbon. Taking Brazilian hydropower project as an example, China – Latin America Productivity Cooperation Investment Fund and Three Gorges Corporation jointly established the Special Purpose Vehicle (SPV) to bid for the investment in the 30 – year franchise rights of two hydropower stations, i.e. Jupia and Ilha Solteria, with a total installed capacity of about 5GW. China – Latin America Capacity Cooperation Fund invested USD 600 million, accounting for 33%, and leverage external financing such as syndicated loan within a short period, enabling SPV to win the bid in time and sign a franchise agreement. At present, the project has entered the post–investment stage, the operation is stable, and the expected return is good.

3. Climate aid to developing countries: China Climate Change South – South Cooperation Fund

In September 2015 during President Xi Jinping's visit to the United States, he officially announced that Chinese Government would contribute 20 billion RMB to establish the Climate Change South – South Cooperation Fund. This fund aims to support other developing countries to respond to climate change, transit to green and low – carbon development, including strengthening the capacity for using Green Climate Fund and the climate adaptability, and strictly control the investment in domestic and foreign high – polluting and heavy – emission projects.

In 2016, China started to carry out the "10 – 100 – 1000" cooperation projects for addressing climate change in developing countries, i.e., 10 low – carbon demonstration areas, 100 climate change mitigation and adaptation projects, and 1000 climate change training quotas. During the 13th Five – Year Plan Period, China will accelerate the establishment of Climate Change South – South Cooperation Fund, coordinate for the promotion of the Belt and Road strategy and international productivity cooperation, and actively help other developing countries implement the sustainable development agenda.

II. Progress of Climate Financing in China

Table 2-6 Status of China Climate Financing

Type		Scale	Period (year)	Description	Data Sources
China Carbon Market	China's pilot carbon trading	Cumulative transaction amount is nearly RMB 2.5 billion	2016	By 31 December 2016, the total transaction amount of pilot carbon markets in seven provinces and cities reached 160 million tons, with a cumulative transaction amount of nearly RMB 2.5 billion, and transaction of carbon allowance online and offline in the secondary markets, which includes Fujian Province, was nearly 64 million tons in 2016, which increased 80% compared with 2015; the transaction amount was about RMB 1.045 billion, which increased nearly 22.1% compared with 2015.	Annual Report of Beijing Carbon Market 2016
	CCER	53 million tons of carbon dioxide e-quivalent	2016	By 31 December 2016, China Certified Emission Reduction Exchange Info – Platform has shown accumulated 2,742 projects approved by CCER publicly, 861 of which have been recorded, and 254 of which have generated 53 million tons of e-mission reductions.	China Certified Emission Reduction Exchange Info – Platform

continued

Type			Scale	Period (year)	Description	Data Sources
Charity fund	Grant	Grant received by China Green Foundation	RMB 54.45 million	2016	Public welfare expenditure in 2016 was RMB 42.43 million.	China Green Foundation 2016 Annual Audit Report
		Amount of domestic and foreign donations received by China used for eco-environment field	RMB 7.66 billion	2015	Eco-environment field received donations of RMB 7.66 billion totally in 2015, which accounted for 6.91% of the total donation and increased 3.41% compared with 2014	Report of China Charity Donation 2015
Traditional financial markets	Traditional international financial markets	Investment in renewable energy by China	RMB 78.3 billion	2016	Compared with 2015, total investment fell by 32%, which is the lowest level since 2013. But China is still the largest investor in the world.	Renewables 2017 Global Status Report
	Domestic financial market	Green credit balance of China major financial institutions in banking sector	RMB 7.51 trillion	By the end of 2016	By the end of 2016, the green credit balance of 21 major financial institutions in banking sector rose to RMB 7.51 trillion, accounting for 8.83% of the loan balance, which is expected to save 188 million tons of standard coal and reduce the emissions of 427 million tons of carbon dioxide equivalent, 2.7146 million tons of chemical oxygen demand, 358,900 tons of ammonia nitrogen, 4.8827 million tons of sulfur dioxide, 2.8269 million tons of nitrogen oxides and save 602 million tons of water.	Report on Social Responsibility of China's Banking Industry in 2016

II. Progress of Climate Financing in China

continued

Type		Scale	Period (year)	Description	Data Sources
Traditional financial markets	Domestic financial market	Energy conservation and environment protection loan balance of China banking sector	RMB 5.81 trillion	2016		Report on Social Responsibility of China's Banking Industry in 2016
		Strategic emerging industry loan balance of China banking sector	RMB 1.70 trillion	2016		
		Energy conservation and environmental protection investment of domestic fiscal funds and national public finance	RMB 474.382 billion	2016	Energy savings of which is RMB 62.265 billion, and renewable energy is RMB 8.612 billion.	Wind Info
		Total amount of green bonds	RMB 238 billion	2016		Report on China Green Bond Market Status in 2016
		Scale of non–labeling green bond issuance	RMB 552 billion	2016		

continued

Type		Scale	Period (year)	Description	Data Sources
Traditional financial markets	Domestic cleaning technology field	Scale of VC / PE investment obtained by China's clean technology industry	RMB 5.328 billion	2016	Compared with 2015, the number of financing cases decreased by 43.88% and the financing scale increased by 91.40%.	Financial data products of China Venture: CVSource
		Scale of completed transaction in China's clean technology industry mergers and acquisitions market	Disclosure transaction size is USD 12.477 billion	2016	The number of completed transaction cases is 156. The completed transaction scale is USD 19.093 billion. Overall, the number and scale of transactions announced by the clean technology mergers and acquisitions market show a downward trend, and the scale of the completed transaction cases has been greatly improved.	
Enterprise direct investment	International market and domestic market	IPO Financing of China's Clean Technology Enterprises	RMB 326 million	2016	Only four companies of clean technology industry issued IPOs in 2016, which decreased 55.56% compared with 2015. The scale of IPO financing was USD 326 million, which fell 87.45% sharply compared with 2015. Among the 4 IPO business, the Shenzhen Stock Exchange listed one and the Hong Kong Stock Exchange listed three.	

II. Progress of Climate Financing in China

continued

Type			Scale	Period (year)	Description	Data Sources
PPP project	Projects in the project warehouse	Ecological construction and environmental conservation	RMB 653.4 billion	2016		PPP Project Database of Ministry of Finance
		Forestry	RMB 8 billion	2016		
	Projects released publicly	Green and low-carbon projects	RMB 5.5 trillion	2016	Among the PPP projects released by PPP comprehensive information platform, there are 6612 green and low carbon projects with investment amount of RMB 5.5 trillion, a year-on-year net increase of 4696 projects with investment amount of RMB 3.53 trillion.	
International multi-bilateral cooperation institutions		Sino-foreign cooperation fund	USD 13.82 billion	2016		Public information sorted by Report Team

III. Climate Financing Process and Cooperation of Other Developing Countries

The rise of emerging markets provides kinetic energy and basic conditions for developing countries to cooperate for addressing climate change, and the climate financing pattern is mainly shifting from the aiding system relying on developed countries to a financing pattern with the coexistence of multiple channels such as the South – South cooperation of developing countries. The main reasons for developing countries to carry out climate change work and South – South cooperation include: (1) the development and increase of the comprehensive national strength and international standing of emerging economies enable them to obtain the capital mechanisms with greater say and autonomy; (2) developing countries are rather sensitive and vulnerable to environmental changes; (3) to achieve some geopolitical objectives; (4) to reduce the emission reduction pressure in the process of international climate negotiations; and (5) to promote the transfer of advanced technology and the output of products etc.

(I) Progress of Climate Financing in Other Developing Countries

1. Climate change policy system has been preliminarily established, while climate financing policy system needs to be improved.

Clear climate goals help set long – term trajectories and provide political signals to businesses and the society.[1] A clear strategic policy

---

[1] Michal Nachmany, etc. The 2015 Global Climate Legislation Study [R/OL]. http://www.lse.ac.uk/GranthamInstitute/wp – content/uploads/2015/05/Global_climate_legislation_study_20151.pdf

III. Climate Financing Process and Cooperation of Other Developing Countries

signal and framework can further enhance investor's confidence, playing a role of guiding and mobilizing private sectors. Developing countries have already started to develop the policies and system framework required to address climate change. According to the research of Grantham Research Institute on Climate Change and the Environment (GRICCE) and Global Legislators Organization (GLO) on global climate legislation in 2015[1], among the 21 African countries included in the research, 16 ones had developed national policy frameworks for addressing climate change. In addition, all African countries except Libya had developed climate related policies to solve energy supply issues.[2] In the Asia-Pacific region and the Latin American Caribbean region, most countries have developed climate change action framework at a national level. It can be said that almost all countries have some forms of climate change legislation, but the climate change policies of most developing country have not yet been transformed into specific action policies, and the systematic climate financing policy system with high correlation is unavailable.

2. Regional finance mechanisms and innovative financial modes have embraced rapid development, gradually indicating their capacity in mobilizing private capital

The international community attaches great importance on the establishment of innovative climate finance platforms, through which regional finance mechanisms could be established in developing countries with innovative financial modes and financial instruments to mobilize the capital from private sectors. Climate and Development Knowledge Network

---

[1] Grantham Research Institute on Climate Change and the Environment (GRICCE) and Global Legislators Organization (GLO) have carried out research on climate legislation in the majority of global greenhouse gas emission countries since 2010. It is currently the most comprehensive climate legislation covering the largest number of countries and has so far issued seven reports.

[2] AfDB. African INDCs: Investment needs and emissions reductions in the energy sector [EB/OL]. http://www.vivideconomics.com/wp-content/uploads/2016/11/AfDB-Report-2016.pdf

(CDKN), Climate Finance Innovation Facility (CFIF), Global Innovation Lab (GIL), etc. are good representatives.

CDKN is a North – South Alliance established through the contribution of British Department for International Development and Dutch Department for Foreign Affairs, and managed by an organization alliance led by Price Waterhouse Coopers. CDKN invested GBP 81,000 to set up a Forest Finance Laboratory at the Southwest Amazon, aiming at attracting capital from public/private sectors to reduce emissions from deforestation and providing capital for the management of sustainable ecological system and the transformation of low – carbon economy.

CFIF was jointly established by the United Nations Environment Programs and the Frankfurt School of Finance & Management for the purpose of supporting financial institutions in developing countries to participate investment in the fields of renewable energy and energy efficiency. Such mechanism provides the financial institutions in developing countries with technical assistance and capital and develops financial products focusing on climate to mobilize more capital to the fields of mitigation and adaptation. Currently, CFIF has supported 15 projects, and the beneficial countries include Pakistan, India, Philippines, Nepal, Cambodia, Mongolia, Singapore, Vietnam and Tonga.

Table 3 – 1 15 Projects supported by CFIF

Projects	Location	Key supporting area
Promotion of Renewable Energy Technologies (RETs) in Pakistan through the establishment of Micro Green Energy Co. (MGEC) supported by microfinancing facility	Pakistan	Renewable Energy, Energy Efficiency
Credit Financing of Green Homes and Solar Off – grid Equipment	India	Renewable Energy, Energy Efficiency
Empowering the Poor through Increasing Access to Renewable/Efficient Energy	Philippines	Renewable Energy

III. Climate Financing Process and Cooperation of Other Developing Countries

continued

Projects	Location	Key supporting area
Taizhou Commercial Bank Energy Small Business Efficiency Lending Programme	China	Renewable Energy, Energy Efficiency
Credit Financing of Solar Home Systems (SHS) for Deprived Communities in Rural Nepal	Nepal	Renewable Energy
Feasibility Study on Access to Financing for RE Appliance for Rural Poor	Cambodia	Renewable Energy
XacBank Enhanced Growth – Development of Carbon Finance Mechanism	Mongolia	Microfinance, Energy Efficiency
Introducing Biogas for Business	Nepal	Renewable Energy
Mitigating and Adapting to Climate Change	Singapore	Energy Efficiency
ESCO Financing and Risk – sharing Programme for Scaling up Energy Performance Contracts in Guangdong Province, China	China	Energy Efficiency
Energy Inclusion Initiative in the Philippines	Philippines	Renewable Energy, Energy Efficiency
Design and Pilot Test of Chamroeun's Clean Energy Loan	Cambodia	Renewable Energy, Energy Efficiency
Supporting rural households of TYM's members in trading with and equipping their houses with solar energy stoves and infrared stoves	Vietnam	Renewable Energy, Energy Efficiency
Capacity Building for Energy Loans in the Pacific (Tonga)	Tonga	Renewable Energy
ESAF Advocacy: Climate Change Mitigation through Clean Energy Products	India	Renewable Energy, Energy Efficiency

GIL was developed jointly by the governments of the UK, US, Denmark, France, Japan, Netherlands, Norway and Germany, a few key development financial institutions and some participants from private sectors, mainly for the innovative demonstration of finance instrument and financing modes. It has attracted USD 170 million since its inception. GIL provides, in the mode of PPP and through innovative financial instruments, specific solutions for the financing challenges faced by projects. In 2015 ~2016, GIL completed the design of the Water Financing Facility (WFF)①, the operation mechanism of which is as follows: Global WFF will be established in the form of a limited liability company, with the initial capital coming from the USD 112 million provided by donors and influential investors, and will provide the equity capital and the first loss guarantee to support the establishment of eight WFFs at national level in developing countries which meet relevant conditions. National WFF will provide long-term low-interest loans for the public or private water resources of the country, and pack them into Asset-Backed Security (ABS). The investment-level bonds priced with local currency will be issued to domestic institutional investors, and the cash flow and residual equity generated from water resource facilities are limited to the repayment of the principal and interest of the bonds. The mechanism is flexible in that apart from the first loss guarantee provided by the global WFF, other multilateral institutions or local governments can also provide additional guarantees or profit commitments for national WFF to further reduce the default risk of investors. When a national WFF can successfully operate and break even, the capital of the global WFF may be taken back and used for investment in the establishment of other national WFFs. It is expected that the leverage ratio of public capital and private capital will reach 1:1.4 through the above method, and it is estimated preliminarily that private capital of USD 1.23 billion could be mobilized every year by 2030. However, a major restriction on the mechanism is the relatively high requirements on monitoring framework, the status of capital market and investors, thus it is difficult to replicate on a large scale. The detailed operation mechanism of WFF is shown in the figure below.

---

① See: https://www.climatefinancelab.org/project/water-finance/

III. Climate Financing Process and Cooperation of Other Developing Countries

Figure 3-1 WFF Operating Mechanism Diagram

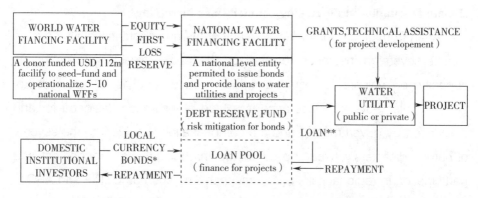

Source: https://www.climatefinancelab.org/project/water-finance/

Case Study III

Kenyan WFF pilot

Kenyan WFF pilot was established in early 2017, and Dutch Department for Foreign Affairs announced to provide a donation of EUR 10 million. The pilot will issue bonds periodically for water resource and public health projects. In addition, seven developing countries relatively meeting conditions are currently selected, and the pilot projects could be duplicated within one or two years. It is expected that Kenyan WFF will issue the first corporate bond at the end of 2017. The preliminary plan of the pilot is to mobilize the local private capital of USD 250 million through the issuance of bonds in the early eight years. According to the quantitative calculation of relevant models, the first loss guarantee of global WFF and the guarantee commitments of governments and other multilateral institutions may cover 42% of the risks of bonds, and mobilize a total public finance of USD 120 million, realizing a leverage ratio of public and private capital of 1:1.3.

3. Risk management tools are developing rapidly and the capital for climate adaptation in developing countries are increasing

Apart from emergency credits, many areas and countries jointly develop climate sovereign insurance to increase the post-disaster reconstruction capacity of vulnerable countries and reduce the fiscal pressure of governments. Sovereign insurance is a type of insurance designed for and purchased by governments, and the compensation is subject to the severity of natural disasters. It may be competed through private arrangement and participation in regional risk sharing mechanism (such as CCRIF and ARC). The advantages of such type of insurance include strict parameters, the stimulation of which will trigger fast compensation. Another advantage is that the cost is fixed and can be estimated accurately. CCRIF is a catastrophe insurance system of developing countries in the Caribbean. When such countries are suffering from natural disasters such as hurricanes and earthquakes, payout could be provided through catastrophe risk insurance funds. The insurance premium of CCRIF is about 50% of commercial insurance market. If the parameter index of a certain year indicates a serious natural disaster, the project will make payout to the insured government according to prior agreements. In terms of reserves, apart from various donations and owned reserve accumulations, it also cooperates with international reinsurance markets through the mode of reinsurance to effectively diversify risks. ARC is a risk control mechanism of African Union established for helping member countries improve the capability of addressing natural disasters such as drought, and its operating mechanism is similar to that of CCRIF.

4. Regional climate funds explore autonomous use of capital, enabling capital to flow to the most needed projects in the region

The integration of capital use and community's priority needs may easily cause investments to focus on short-term goals and has the risk of contradicting the national long-term climate policy goals. To realize the

III. Climate Financing Process and Cooperation of Other Developing Countries

balance between the two is a major challenge for fund operators and stakeholders. The finance mechanism established in developing countries have started to include local demands into capital use plans, and keep projects in line with national long-term strategic goals. Ethiopian Climate Resilient Green Economy (CRGE) and Kenyan County Climate Change Funds (CCCFs) are beneficial attempts in this field. The CCCFs in Wajir and Makueni are trial pilots of Kenyan Administration for Drought authority, and they are established with the technical support of the adaptation association and the donation of GBP 6.5 million provided by British Department to Kenya government for supporting such projects. The pilot aims to promote the mainstreaming of climate change in county-level development planning, strengthen the right of county governments to use climate adaption capital, and give priority to the consideration of community demands. Each county will allocate 20% of the fund for capacity building to support the water source governance led by communities. In Wajir, the fund has invested 12 district projects and 2 county projects, with a total investment of USD 480,756, which are used for improving the capacity of water users to use water sources in a sustainable manner.[1]

(II) Developing Countries Actively Conducting Climate Aid and Cooperation

In recent years, developing countries have not only actively responded to climate change by themselves, but also started to provide external climate assistance. The amount of foreign aid from China, the largest developing country, grew rapidly over the past decades. The amount of aid China provided to Africa has exceeded that of the World Bank. From 2001 to 2013 the amount of China's foreign aid rose from USD 743 million to USD 7.462 billion, with an average annual growth rate of 21.20%. China's

---

[1] ODI. Decentralising Climate Finance – Insights from Kenya and Ethiopia[EB/OL]. https://www.odi.org/sites/odi.org.uk/files/resource-documents/11804.pdf

international aid growth rate is equivalent to 2 times of the world average growth rate (10.22%), 2.28 times that of the United States (9.30%), 3.96 times that of Japan (5.35%), 2.96 times that of the UK (7.17%), 1.72 times that of Germany (12.32%)①. As a branch of China's foreign aid, the climate aid undoubtedly shows a significant increase in recent years as well.

Other developing countries also strengthen cooperation in climate financing through cooperation in creating multilateral institutions or dialogue platforms. Climate Policy Initiative (CPI) tracking data of 2015/2016 indicates that USD 3 billion/year on average of climate finance flowing from developing countries to developed countries and USD 8 billion of flows between different developing countries②. Among the more prominent multilateral institutions established by developing countries are the New Development Bank (former BRICS Development Bank), the African Development Bank, the Inter-American Development Bank and the Asian Infrastructure Investment Bank. Such multilateral agencies directly assist specific projects or local implementing entities through subsidies, loans, debts, stocks, credit lines, etc. In addition, such multilateral institutions also realize the goal of climate aid indirectly by supporting the national policies of recipient countries, for example, providing technical assistance for some countries to develop policies related to climate change, or promoting environmental issues to become the mainstream factor of national general strategic planning. In addition, developing countries began to fund to international multilateral institutions led by developed countries, mainly providing funds to the Global Environment Facility (GEF) and the Green Climate Fund (GCF). Although the overall contribution ratio was less than

---

① HU Angang, ZHANG Junyi, GAO Yuning. Foreign aid and national soft power – China's Current Situation and Countermeasures [J]. 2017, 70(3). Wuhan: Journal of Wuhan University, 2017.

② CPI. The Global Landscape of Climate Finance 2017 [R/OL]. https://climatepolicyinitiative. org/wp-content/uploads/2017/10/2017-Global-Landscape-of-Climate-Finance.pdf

III. Climate Financing Process and Cooperation of Other Developing Countries

1% in the total amount received by GEF and GCF①, climate finance from developing countries is very symbolic and meaningful.

(Ⅲ) Demands of Developing Countries and China's Countermeasures

1. Developing countries are facing double challenges of capital shortage and capacity building

Though developing countries have made great achievements in dealing with climate change, they still face huge challenges. India needs climate finance of USD 200 billion to realize Intended Nationally Determined Contributions (INDC) goal②. Among the 53 African countries' INDCs, the needs of 43 countries will reach USD 1150 billion by 2030, and it is predicted that 82% of the capital for mitigation in 23 countries comes from the international community. In terms of industry level, the energy sector has the second largest needs accounting for 37% (USD 159 billion), the agriculture, forestry and other land use sector accounts for 42%, and waste treatment sector accounts for 20%.③

According to the survey made by St George Regional Collaboration Center (SGRCC) on INDC plans,④ Africa is in urgent need of capacity building, including increasing the capacity of governments in constructing the legal and policy frameworks that may promote and stimulate private sectors to participate in climate financing. Although international climate funds offer many opportunities for accessing climate finance, generally African countries have

---

① See official website of GEF and GCF

② See:http://www.eurasiareview.com/30082017-climate-finance-in-india-a-case-of-policy-paralysis-analysis/

③ AfDB. African INDCs: Investment needs and emissions reductions in the energy sector [EB/OL]. http://www.vivideconomics.com/wp-content/uploads/2016/11/AfDB-Report-2016.pdf

④ St George Regional Collaboration Center. Report on the nationally determined contributions survey conducted by the Nairobi Framework Partnership in 2016[EB/OL]. http://unfccc.int/files/secretariat/regional_collaboration_centres/application/pdf/report_ndc_survey_final.pdf

difficulty in designing the proposals to meet climate funds' requirements, thus being unable to make full use of existing funding opportunities①. The focus of East Africa is to implement specific projects/programs to develop a solution containing diversified clean energy technologies, which will guarantee the general state energy security through strengthening regional availability, affordability and reliability.② The experience of Ethiopia and Kenya also indicates that local governments have no sufficient technical expertise to support local project design and implementation, which slows down the development of projects, impedes the localization of international climate capital in developing countries, and is not good for capital to flow to project fields and regions in urgent need. The transportation, waste treatment and agriculture in Asia Pacific region are the fields in most urgent need of support to achieve the long-term INDC goals. Latin American countries focus more on how to stimulate investment from private sectors and how to establish a MRV system.

2. The environmental carrying capacity of the countries along "the Belt and Road" is extremely fragile and it is required green investment.

Since 2013, when President Xi Jinping proposed "the Belt and Road" strategy, China has been actively promoting the economic and trading cooperation with the involved countries, strengthening exchange and dialogue with the involved countries. "The Belt and Road" is currently the longest economic corridor in the world, involving over 60 countries in Europe and Asia, mainly in Central Asia and Southeast Asia. Generally, such countries have the problems of low opening-up degree, backward infrastructure construction and low social and economic development level. Specifically, Central Asia is featured by large area of desert and lack of

---

① See: http://www.linkedin.com/pulse/africa-needs-climate-finance-capacity-access-funds-critical-domfeh/

② St George Regional Collaboration Center. Report on the nationally determined contributions survey conducted by the Nairobi Framework Partnership in 2016[EB/OL].

III. Climate Financing Process and Cooperation of Other Developing Countries

water resources, Southeast Asia is experiencing the rapid shrinkage of tropical rainforest area, and South Asia is suffering from heavy pollution to marine resources due to massive sewage drainage by coastal cities.① Among such countries along the Belt and Road, over half have relatively extensive economic development modes, with the energy consumption per unit GDP and the emissions of carbon dioxide exceeding 1.5 times of the average world level.② Therefore, the environmental protection situation in these countries is extremely pessimistic, and the need for green investment is very high. China should take this as an important opportunity to drive the green investment in the countries involved, which can not only help such countries perform transformation of economic structure to realize sustainable development, but also promote the technical improvement and equipment optimization of domestic relevant industries, forcing enterprises to speed up research and development and activating their innovation capacity.③

---

① YANG Zhen, SHEN Enwei. Study on Accelerating the Green Investment Along the Belt and Road under "the Belt and Road" Strategy[J]. 2016 (9):21 – 24. Foreign Economic and Trade Practices.

② WANG Zhongping. Combating Climate Change, The importance of "One Belt One Road" [N/OL]. http://www.crntt.com/doc/1046/6/6/4/104666452.html? coluid =7&kindid =0&docid = 104666452

③ YANG Zhen, SHEN Enwei. Study on Accelerating the Green Investment Along the Belt and Road under "the Belt and Road" Strategy[J]. 2016 (9):21 – 24. Foreign Economic and Trade Practices.

IV. Trump Presidency's Impact on Global Climate Financing

Since his inauguration in January 2017, President Donald J. Trump has dramatically changed US energy and climate change policies of Barack Obama. Amongst the policy changes, the most significant ones include: cancelling "Climate Change Action", releasing "American First Energy Plan", announcing withdrawal of the *Paris Agreement*, dramatically cutting down climate policy and research related budgets, placing limits on Environmental Protection Agency, the lifting of restrictions on fossil fuel production and regulations on energy sector, cancelling payments to UN climate change programs. For specific policy contents and release time, please refer to Table 4 – 1.

Table 4 – 1 Trump policies and actions that have significant impact on global climate governance and climate financing

Time	Main Policies or Actions
21 January 2017	Cancel "Climate Action Plan" and issue "American First Energy Plan"
16 March 2017	"America First Budget Plan" and "Presidential Executive Order promoting Energy Independence and Economic Growth" dramatically cut budgets on climate policy and research; cease payments to GCF and CIFs, as well as reduce bilateral climate aids
1 June 2017	Announce to withdraw from the *Paris Agreement*
8 June 2017	Pass the "Financial Choice Act", investors are no longer able to require companies to disclose the climate risk information through the shareholder proposal process

The US local large – scale enterprises, investors, state and municipal governments, universities and Think – Tanks have already taken down –

IV. Trump Presidency's Impact on Global Climate Financing

side up actions towards Trump's policies. But a series of simulation researches indicate that there has already been high risk US cannot achieve the goal of reducing emissions by 26% ~28% on 2005 level by 2025. As a result, it may cause domino effect and lead other nations to erode their goals, harming the global target of the Pairs Agreement①. In addition, Trump has clearly indicated that the US will no longer make any contributions to the Green Climate Fund and Climate Investment Funds②. Meanwhile, measures such as reducing bilateral climate aids will also have profound influence on global climate financing pattern.

(I) Influence on Global Climate Capital Flow

1. Overview of American contributions to global climate financing during the Obama Presidency

In Obama presidency the US provided developing countries with financial aids for addressing climate change through various channels such as multilateral development banks, Export – Import Bank of the United States, US Overseas Private Investment Corporation and climate funds.

For the period of 2010 to 2015, the US public international climate finance assistance and contribution reached USD 15.6 billion. This includes bilateral assistance, MDBs, development assistance and official export credit. ③

According to UNFCCC statistics, the US public international climate finance was USD 4.835 billion and USD 5.1 billion respectively in 2013 and

---

① Climate Initiative & MIT. Analysis: U. S. Role in the Paris Agreement [N/OL]. https://www. climateinteractive. org/analysis/us – role – in – paris/

② Switzer Foundation. Green Finance: The next frontier of US – China Climate Co – operation [N/OL]. https://www. switzernetwork. org/switzer – fellow – thought – leadership/green – finance – next – frontier – us – china – climate – cooperation

③ Switzer Foundation. Green Finance: The next frontier of US – China Climate Co – operation [N/OL]. https://www. switzernetwork. org/switzer – fellow – thought – leadership/green – finance – next – frontier – us – china – climate – cooperation

2014.① Besides, US contributed USD 2 million to the Climate Technology Center and Network which is under UNFCCC framework,② and pledged USD 3 billion towards the Green Climate Fund although Obama only provided 1 billion before he left the office. In terms of country distribution, 42% flowed to Asia, 36% to Africa, 19% to Latin America and the Caribbean, and the rest to developing economies in Europe and the Middle East.③

In 2014, developed countries provided total public climate finance of USD 43.2 billion to developing countries④, with the US contributing USD 5.1 billion, accounting for 11.8% of the total amount. It is additionally important to note that public finance flows have a catalytic role on private investors and consequently the private capital mobilized from developed to developing countries in 2014 amounted to USD 16.7 billion according to OECD⑤. This in turn allows us to calculate that for each public dollar spent, USD 0.38 of private capital is mobilized. There is yet no statistical data on the amount of private funds mobilized by the US. Assuming that the ability of developed country governments to mobilize private capital is approximately equal across countries, then the US mobilized private capital is approximately USD 1.938 billion. The estimated total amount of climate

---

① U.S. Department of State. 2016 second biennial reports of United States under the UNFCCC[EB/OL]. https://unfccc.int/files/national_reports/biennial_reports_and_iar/submitted_biennial_reports/application/pdf/2016_second_biennial_report_of_the_united_states_.pdf

② U.S. Department of State. 2016 second biennial reports of United States under the UNFCCC[EB/OL]. https://unfccc.int/files/national_reports/biennial_reports_and_iar/submitted_biennial_reports/application/pdf/2016_second_biennial_report_of_the_united_states_.pdf

③ U.S. Department of State. 2016 second biennial reports of United States under the UNFCCC[EB/OL]. https://unfccc.int/files/national_reports/biennial_reports_and_iar/submitted_biennial_reports/application/pdf/2016_second_biennial_report_of_the_united_states_.pdf

④ UNFCCC. 2016 biennial assessment and overview of climate finance flows report[EB/OL]. http://unfccc.int/files/cooperation_and_support/financial_mechanism/standing_committee/application/pdf/2016_ba_summary_and_recommendations.pdf

⑤ OECD, CPI. Climate Finance? in 2013-14 and the USD 100 billion goal [EB/OL]. http://www.oecd-ilibrary.org/docserver/download/9715381e.pdf? expires = 1517901239&id = id&accname = guest&checksum = 75A4697E7FA9ADF080FE02BD7169CB4B

IV. Trump Presidency's Impact on Global Climate Financing

finance provided by the US to developing countries in 2014 is therefore USD 7.038 billion.

2. The Trump Administration slashed climate funding support

The Trump Administration cut the budget of US Environmental Protection Agency by 31.4% in the Fiscal Year 2018 Budget, and the "Clean Power Plan", International Climate Change projects and Climate Change Research and Cooperation projects implemented by the Environmental Protection Agency will not get financial support any more. The budget of US National Aeronautics and Space Administration is reduced from USD 19.3 billion to USD 19.1 billion, where the reduction of the budget for geoscientific research is USD 102 million and almost all reductions are for the climate change research projects. The ocean management, research and education fund of USD 250 million of the US National Oceanic and Atmospheric Administration is canceled. Moreover, the budget of US Agency for International Development is reduced by 28%. ①

In addition to the budget cuts of funds for basic science research in the United States, the White House also stated intentions to cancel a number of overseas aiding funds and cut contribution to the UN by as much as 50%.

Based on the White House's budget paper the New York Times has analyzed the implications by budget items such as MDBs, development assistance, US Agency for International Development (USAID) operation, and other development and humanitarian assistance. The White House's budget proposal shows that the MDB allocations will be reduced by 42.1%, the allocation of development aid will be cut by 79.6%, the USAID budget will be cut by 27.1% and the other development and humanitarian aid budgets will be cut by 82.7%. ② Based on the budget it is not possible

---

① U.S. Government. Budget of the U.S. Government. A new foundation for American Greatness [EB/OL]. https://www.whitehouse.gov/sites/whitehouse.gov/files/omb/budget/fy2018/budget.pdf

② New York Times. Trump Budget Details [N/OL]. https://www.nytimes.com/interactive/2017/05/23/us/politics/trump-budget-details.html?mcubz=3

to specifically allocate reductions to climate finance, but assuming that the aid funds are evenly distributed to various projects, it is estimated that the US's reduction on climate financing through multilateral development banks, development assistance, USAID, and other development and humanitarian assistance will be approximately 42.1%, 79.6%, 27.1% and 82.7% respectively. As the Trump Administration limits international finance towards items critical to its perceived national security, it is consequently proportionally more skeptical of climate change funding than of other international issues, and therefore this estimate is more likely to be an under – rather than an overestimation. In addition, Trump has clearly claimed that the US will no longer provide funding towards the Green Climate Fund, which means that the rest USD 2 billion promised by Obama is no longer guaranteed.

If the world sticks to the UNFCCC and the commitment to addressing climate change, the financial gap due to the fiscal budget reduction by the US could be partially filled by alternative channels, which will affect the global climate treatment pattern and the role of China. In fact, the role played by the US in the global climate finance architecture has always been relatively limited while China's foreign assistance (or ODA) has gradually increased proportionally within the total global foreign assistance (total global ODA) over the past few years. Today, China has become the world's fourth largest foreign aid provider[1]. Although the European financial sector and climate think tanks have been making ongoing efforts in rebuilding the climate finance architecture, mobilizing private finance, and promoting the innovation of climate finance products and instruments, the climate financing hot –spots have turned to China, India and other Asian regions. Consequently, the global climate leadership is entering a transitional era.

---

[1] HU Angang, ZHANG Junyi, GAO Yuning. External Aids and National Soft Power – China's Current Situation and Countermeasures [J]. 2017, 70(3). Wuhan: Journal of Wuhan University, 2017.

IV. Trump Presidency's Impact on Global Climate Financing

(II) Changes to Global Climate Governance Pattern

1. Climate financing leadership is entering into a transitional era

The announcement of exiting *Paris Agreement* by the Trump Administration means that the US is giving up its leadership in global climate governance. Meanwhile, the European Union is struggling to cope with a series of problems such as Brexit and refugee flows, they are willing, but largely powerless in addressing climate issues[①]. Therefore, the current international community holds expectations for China to fill the leadership vacuum left by the US.

The withdrawal of the US from the *Paris Agreement*, may remind people of the chain effects caused by the US exit from the Kyoto Protocol in 2001 by President Bush. At that time, many developed countries followed the examples of US or reduced the execution power of emission reduction, causing block of the implementation of Kyoto Protocol. But many experts hold that it is unlikely the *Paris Agreement* would follow the footsteps of the Kyoto Protocol.[②] In fact, the exit of the US does not maul the determination of other contracted countries to address climate change as concerned by all other parties. Instead, after Trump's announcement of withdrawal, many world leaders condemned it and reaffirmed their determination to implement the *Paris Agreement*. The strong and prompt active response from the international community indicates the resilience of global climate change system established in the past two decades. The real damcgy caused by the exit of US from the Paris Agreement may he its own leadership in international affairs. This was indicated most distinctly at the G20 meeting in

---

① Caixin. com. The exit of the US is a huge loss to the *Paris Agreement*, Whether China can take on the global leadership vacuum? [N/OL]. http://china.caixin.com/2017-06-02/101097119.html

② Xinhua News Agency. What is the difference between Trump's exist from the *Paris Agreement* and George W Bush's exit from the Kyoto Protocol? [N/OL]. http://hunan.voc.com.cn/xhn/article/201706/201706061729545724.html

Hamburg in July 2017①, where Trump was isolated with a 19 – 1 split on climate change, towards which Angela Merkel 'deplored' the US decision to withdraw②. The decline of American leadership is reflected not only in climate field, but also in other international affairs. For example, Trump questioned the role of the US in NATO, exit the trans – Pacific partnership, and substantially cut the financial support to UN programs. This will further aggravate the decline of the leadership of the US in global climate financing.

Overall, the withdrawal of the US will not shake the foundations of the global climate governance system as concerned by many people; instead, it may prompt more active actions. The leadership in climate change governance is at a transitional phase, with the geographical focus gradually shifting to Asian region such as China and India, while the US still maintains the strength in new energy technologies, strong financing systems and financial market vitality. It is difficult to form a single core leadership for climate financing in the short term.

An early indication of the potential effect on the global governance of climate finance is the proliferation of climate related initiatives within and between developing countries. These efforts include the incorporation of climate finance efforts into regional initiatives such as ASEAN and the African Union as well as clear climate finance ambitions in developing country – led MDBs and funds such as the New Development Bank, the Asian Infrastructure Investment Bank and the Silk Road Fund. In total, this is an indication of greater autonomy for developing countries within climate change governance. Symbolic of this change is the shift in approach from Kyoto to Paris within the UNFCCC negotiations, in which the prior was

---

① Michele M. Betsill. Trump's Paris withdrawal and the reconfiguration of global climate change governance[J]. Chinese Journal of Population Resources and Environment.

② The Guardian. G20 summit: 'G19' leave Trump alone in joint statement on climate change – as it happened [N/OL]. https://www. theguardian. com/world/live/2017/jul/08/g20 – summit – may – meets – world – leaders – in – bid – to – boost – brexit – trade – prospects

IV. Trump Presidency's Impact on Global Climate Financing

centered around a top-down approach of distributing developed countries contributions, whereas the latter was based on a bottom-up approach of all countries' contributions allowing developing countries a clearer autonomy and channel of involvement.

2. The global investment trend towards new energy transformation has not been significantly affected

Investment in the new energy sector has risen rapidly in the past few years. Relevant data show that even during the global recession period from 2008 to 2012, the growth rate of investment in the new energy sector was higher than that of other sectors.[①] The Trump administration's new policy will be difficult to reverse this trend, mainly because investors' interest in low-carbon and new energy market has not been frustrated.

Investments in new energy sectors have risen rapidly in the past few years, partly thanks to the dramatic decline of clean energy cost. For example, the cost of wind power dropped from USD 101 ~ 169 per megawatt in 2009 to USD 32 ~62 per megawatt in 2016, decreasing by 66%, and the cost of solar power dropped from USD 323 ~ 394 per megawatt in 2009 to USD 49 ~61 in 2016, decreasing by 85%.

Globally, we find that after President Trump announced to withdraw from the *Paris Agreement*, VIX volatility (panic) index dropped from 10.20 to 9.70, indicating that the final decision on exiting the *Paris Agreement* has ended the conjecture. The Euro was flat against the Dollar. The rising momentum of MSCI World Low Carbon Target index and Wilder Hill New Energy Global Innovation Index (NEX) since the beginning of 2017 has not been affected. The Stowe Global Coal index dropped slightly during following the announcement, but it didn't take long to rebound and has stopped the decline since April. S&P 500 oil, gas exploration and production sub industry GICS level, however, has not been boosted by Trump's

---

[①] Joseph P. Tomain. A US Clean Energy Transition and the Trump Administration[J]. Social Science Electronic Publishing.

"America First Energy Plan" and the decision to withdraw from the *Paris Agreement and has maintained the downward trend since the beginning of* 2017.

From the performance of China's new energy and traditional energy sector in the stock market, we can see that clean energy stocks outperformed the market and fossil fuel stocks within the few days following the "big" announcement. In the longer term, China's low carbon index went up from 4508.5733 on January 3, 2017 to 5567.1042 as of December 29, 2017, a 23.47% increase; the solar power index kept declining from March 2017, but climbed steadily by 6.6% since June after the "Withdrawal" decision was announced, indicating the announcement quelled the market's suspicious emotion. Overall, it seems that low-carbon investors' interest in the clean energy sector has not been affected by a series of "de-climate" policies issued by President Trump.

Figure 4-1 Performance of China's major stock index from January to December 2017

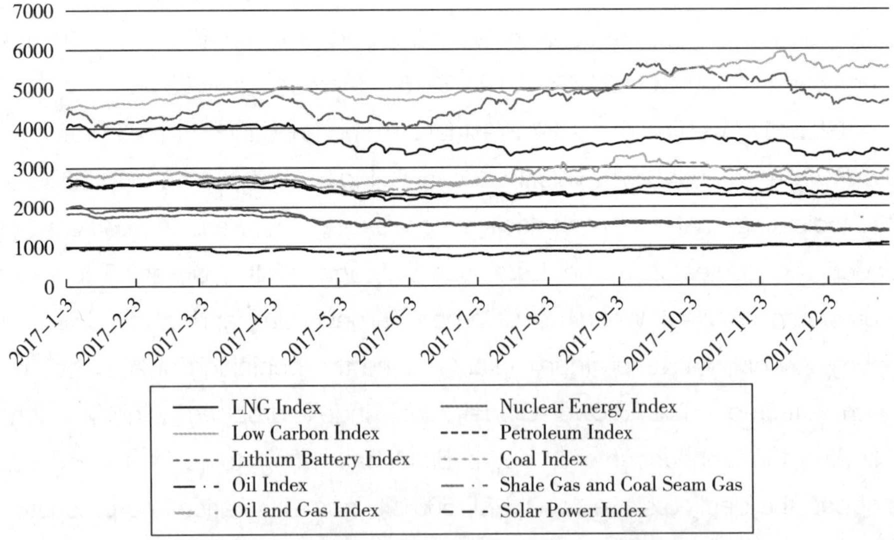

Source: Wind.

IV. Trump Presidency's Impact on Global Climate Financing

3. The withdrawal of the US would squeeze China's emission space and increase China's emission reduction costs

A series of simulation research has already shown that there is high risk that the United States cannot achieve its emission reduction goal of 26 −28% below the 2005 levels by 2025. According to Dai Han −Cheng's latest study, the US withdrawal from the *Paris Agreement* will squeeze China's emission space and increase China's emission reduction costs. Specifically, under the 2 −degree target, if the US only reduces its emissions by 20%, 13% and 0% below the 2005 levels by 2025, it will decrease CO_2 emissions space by 1.7%, 2.8% and 5.0% in China, and the carbon price will rise by USD 4.4 ~14.6 /ton, the additional GDP loss will be USD 21.98 ~71.1 billion in China by 2030[1]. Yet, there is also positive aspect, US's retreat may reinforce China's existing dominance in the clean energy sector[2].

The international community holds high expectation that China will fill the global climate leadership vacuum left by the US. Some scholars have suggested that China shall rebuild global shared leadership by replacing the Sino − US G2 partnership with a Climate 5 (C5) partnership that includes China, the EU, India, Brazil and South Africa.[3]

(Ⅲ) **New Role of China in Global Climate Financing**

The reduction of the US's climate change efforts coincides with China's expansion, providing a new role of China. Whereas China is often considered one of the main stumbling blocks in the failure of the COP 16 Copenhagen negotiations, it is widely acclaimed as a critical driver of the success of the

---

[1] DAI Hancheng, ZHANG Haibin, WANG Wentao. The impacts of U.S. withdrawal from the Paris Agreement on the carbon emission space and mitigation cost of China, EU, and Japan under the constraints of the global carbon emission space[J]. 2017,8(4):226 −234. Advance in Climate Change Research.

[2] ZHANG Haibin, DAI Hancheng, LAI Huaxia, WANG Wentao. US withdraw from Paris[J]. 2017,1(6). Advance in Climate Change Research.

[3] ZHANG Haibin, DAI Hancheng, LAI Huaxia, WANG Wentao. US withdraw from Paris[J]. 2017,1(6). Advance in Climate Change Research.

COP 21 Paris negotiations. Within this limited time period China substantially increased its climate change ambitions domestically and abroad. Domestically, China has formalized the concept and pursuit of an ecological civilization, including environmental sustainability as a cross – cutting theme of public policy. In addition, since 2016 when seven ministries and commissions issued the *Guidelines on Establishing a Green Financial System*, China accelerated the process of green financial development. Internationally, China has become an upholder of the Paris *Agreement* and supporter of globally coordinated efforts against climate change. In the opening report of the 19th CPC National Congress, General Secretary Xi Jinping pointed out that "(China) leads the international cooperation on climate change, and becomes an important participant, contributor and leader of global ecological civilization". He further stressed to "stick to be environment – friendly, cooperate to address climate change, and protect the earth on which human lives".

The trend of China's new role within climate change culminated with President Xi Jinping's speech in Davos at the World Economic Forum's 2017 session. Within the implicit context of the demise of the US leadership of globalization and substantial anti – globalization sentiment across developed countries, President Xi Jinping advocated for joint efforts against climate change and highlighted the mutual benefits of the globalization of trade, finance, and governance. With such commitment from the Chinese side, China and the EU currently spearhead the global climate change efforts jointly. This was solidified in the rhetoric coming out of the EU – China summit as voiced by the President of the European Commission: "As far as the European side is concerned, we are happy to see that China is agreeing to our unhappiness about the American climate decision. This is helpful, this is responsible, and this is about inviting both, China and the European Union, to proceed with the implementation of the Paris *Agreement*". ①

---

① European Commission. Press Release: EU – China Summit: Moving forward with our global partnership[N/OL]. http://europa.eu/rapid/press – release_IP – 17 – 1524_en.htm

IV. Trump Presidency's Impact on Global Climate Financing

Within the narrower scope of climate finance specifically, China has launched and participates in a wide range of financing initiatives. The primary channels of China's public annual international climate financing include MDBs, funds, the direct involvement of Chinese policy banks, and Chinese state – owned enterprises abroad. Firstly, in terms of MDBs, China's role includes contributing to the establishment of the New Development Bank, the Asian Infrastructure Investment Bank, and the intention to establish a Shanghai Development Cooperation (SDC) bank[①], which all have clear ambitions to contribute to climate financing. Secondly, in terms of sovereign backed funds, China has established and works as primary financier of 14 international funds that all prioritize climate finance, such as the Silk Road Fund, the China –LAC Industrial Cooperation Investment Fund, and the South –South Climate Change Fund. Thirdly, Chinese policy banks, as well as state owned financial and other enterprises finance, construct, and operate climate related projects abroad. The data suggest that the China Development Bank and China Exim Bank combined have financed USD 32 billion worth of hydropower plants abroad since 2007[②].

[①] Xinhua. Shanghai Cooperation Organization prime ministers meet in Bishkek[N/OL]. http://news.xinhuanet.com/english/2016 –11/03/c_135803540.htm

[②] Financial Times. China poised to lead on green finance at G –20 meeting [N/OL]. https://www.ft.com/content/65ff1e54 –ec78 –380e –ac10 –5a633eea983b

V. Cooperation of Multilateral Development Banks

(I) High Potential of Cooperation among Multilateral Development Banks

Satisfying the need for infrastructure investment is one of greatest current challenges of international finance. An often – cited figure by the Asian Development Bank (ADB), estimates that Asia alone requires USD 1.7 trillion of annual infrastructure investment in the period until 2030[①]. Discussing how to meet global development finance needs, the 2015 United Nations International Conference on Finance for Development in Addis Ababa highlighted the role of multilateral development banks (MDBs) and called for establishing global fora to enhance multilateral finance cooperation[②]. The agenda highlighted the importance of MDBs in terms of finance and know – how, low volatility & counter – cyclical lending, long –term lending, concessional lending, and mobilizing financial markets. All these constitute important components of reaching the Sustainable Development Goals[③].

MDBs can cooperate in a number of ways across financial and technical mechanisms, social and industrial sectors, and geographic regions. Furthermore, their cooperation functions as standard – and agenda setting

---

[①] ADB. Meeting Asia's Infrastructure Needs[EB/OL]. https://www.adb.org/sites/default/files/publication/227496/special – report – infrastructure.pdf

[②] UN. Addis Ababa Action Agenda [EB/OL]. http://www.un.org/esa/ffd/wp – content/uploads/2015/08/AAAA_Outcome.pdf

[③] UN. Addis Ababa Action Agenda[EB/OL]. http://www.un.org/esa/ffd/wp – content/uploads/2015/08/AAAA_Outcome.pdf

V. Cooperation of Multilateral Development Banks

within technical standards of development finance, to social and environmental safeguards, as well as towards the usage various financing tools. A recent development is the proliferation of new MDBs. In addition to the New Development Bank by the BRICS countries, the AIIB has been established, and the Shanghai Cooperation Organization is setting up an MDB to support its members. It is worth noting that all of these have Chinese involvement or leadership. The modus operandi of these MDBs supports the current global financial system rather than aims to change it, and consequently operate in parallel and with overlap towards existing MDBs, thus providing a clear scope for cooperation.

When it comes to MDB cooperation and green finance, international climate finance provided by MDBs is a critical tool for turning the world's lofty promises of the COP 21 into a tangible green transition on the ground. As actors in this process, the requirements for green spending of MDBs are increasing, while the number of climate related funds are proliferating. As this is a general trend across MDBs, there is great potential for their cooperation, providing both mutual organizational benefit and greater environmental impact. Concretely, their cooperation can increase the overall impact through the efficiency by scale of shared financial resources, combined technical expertise, and joint access and legitimacy in local environments.

As the EIB expands its geographic scope and the AIIB has become operational, cooperation between the two has become a practical possibility. Simultaneously, since the EIB is increasingly focusing on green investments and the AIIB is green in nature, their engagements within green finance provide a clear overlap. While, on 30May 2016 the AIIB and EIB signed an MoU stating their intent to cooperate, they have yet to realize this intention. Their only direct point of interaction was in the AIIB and World Bank Group (WBG) co-financed Trans Anatolian Natural Gas Pipeline Project (TANAP), which was supported with loans from a consortium of MDBs including the Asian Development Bank (ADB), European Bank of Reconstruction and

Development (EBRD), as well as the EIB. On their potential cooperation, Jin Liqun, the President of AIIB, has stated: "The agreement is crucial to expand our partnership in addressing the monumental infrastructure financing needs around the world". Additionally, Werner Hoyer, the President of the EIB stated: "... looking forward to working together with the AIIB to address global challenges, in particular programs that would help tackle climate change, ensuring the provision of sustainable transport and providing clean water". ①

The cooperation framework is based on three analytical approaches to analyze all possible cooperation aspects between the AIIB and EIB. Then it goes in depth with the concrete potential within green finance. The logic of this process is that a general assessment of cooperation possibilities provides the prerequisite basis for assessing green finance cooperation. The three analytical approaches are: 1. Organizational mandates, 2. operational capabilities, 3. political circumstances. As shown in Figure 5 – 1, MDB cooperation can be framed in terms of three categories. First, all possible mechanisms are listed in the left column separated into financial, technical, and intra – bank cooperation. Second, the middle column lists all possible sectors for cooperation, categorized based on project categorizations of the EIB, World Bank Group, and Overseas Development Institute. These sectors are intentionally overlapping but are categorized to encompass the most relevant perspectives on sectorial categories. Third, the right column lists regions and countries. Any project of MDB cooperation has to fit as least one and possibly more items in each of the three categories, e.g. advisory services (mechanism) for energy infrastructure (sector) in Pakistan (geography).

---

① China Daily. AIIB, EIB sign a pact to help finance infrastructure [EB/OL]. http://www.chinadaily.com.cn/business/2016 – 05/31/content_25544988.htm

V. Cooperation of Multilateral Development Banks

Figure 5 – 1   MDBs Cooperation Framework

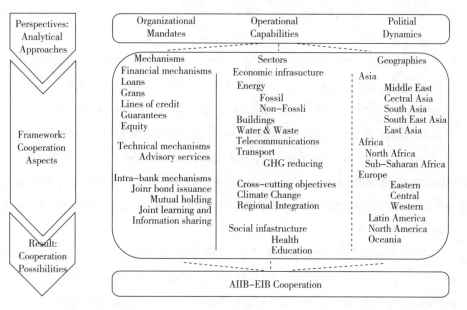

(Ⅱ) Cooperation between AIIB and EIB

1. Asian Infrastructure Investment Bank

Established at the initiative of China, the AIIB has 50 founding members of primarily Asian and European origin, altogether providing capital of USD 100 billion. While being a developing country led MDB, it is established on the principle of working within the existing global financial regime. It is further based on the principles of being lean, clean and green. In terms of governance, financial contributions and voting rights are closely correlated, with 78% of voting rights for Asian member countries. While not having a formal veto power, China has 27% of voting rights providing a clear leadership as well as de facto veto power in decisions that require a supermajority of votes of 75%. However, as membership increases, this veto power will be lost eventually, emphasizing its nature as a democratic institution, as expressed by the AIIB President Jin Liqun①.

Similar to other MDBs, the AIIB is expected to issue debt in terms of bonds for investors to purchase on international financial markets. However, due to its sufficiently large starting capital and only gradual expansion of investments, this has not yet taken place despite achieve an AAA rating by Moody's in June 2017. The thematic priorities of the AIIB have three components: Sustainable infrastructure to assist developing countries reach their environmental and development goals, cross – country connectivity infrastructure across a broad range of sectors across Asia and beyond, and private capital mobilization in partnership with other MDBs, governments and private financiers.

As released in June 2017, the AIIB energy strategy aims to provide electricity to millions of people in Asia while simultaneously assisting countries in reaching their environmental commitments②. It will do this both through renewable energy projects, investments in energy efficiency, rehabilitation and upgrading of existing plants, and transmission and distribution networks. It specifically highlights that it will cooperate with other MDBs in these efforts. However, it makes no quantitative proportionate commitment to financing labeled as "green", even though the term is part of its core principles.

2. European Investment Bank

The EIB functions as the EU's non – profit lending institution and was established in 1958 under the Treaty of Rome. With a total subscribed capital of EUR 243 billion it is, by a small margin to the WBG, the world's largest MDB, and represents the interest of the EU's members in financing sustainable development projects inside and outside the union. Through a variety of financial mechanisms, it prioritizes financing for projects in

---

① China Daily. AIIB Chief rules out China veto power [N/OL]. http://www.chinadaily.com.cn/business/2016 -01/27/content_23265846.htm

② AIIB. Energy Sector Strategy: Sustainable Energy for Asia [EB/OL]. https://www.aiib.org/en/policies -strategies/strategies/.content/index/Energy -Strategy -Discussion -Draft.pdf

V. Cooperation of Multilateral Development Banks

infrastructure, innovation, and climate change mitigation and adaptation. While the EIB is managed through the EU Commission, the organizational governance and structure plays a role in the functions of the Bank. As an EU institution, its strategic goals are aligned with those of other EU institutions, and it therefore works closely with these to support each other and avoid overlaps. [1]

In order for the EIB to maximize its lending, it borrows on global financial markets. The bulk of its funding comes from bond issuance on international capital markets. These bonds are both bought by institutional investors, such as from the EU and other countries, but also by retail investors internationally[2]. The EIB carries out its mandate by lending, blending, and advising. Lending accounts for 90% of the EIB's total financial commitments. For large scale projects (above USD 30 million), the EIB can finance 50% of a project, but on average the number is about 30%[3]. This is a common method of financing for international financial institutions (IFIs). Blending, in the EIB's own terms, is a variety of sophisticated finance tools that go beyond conventional financing through loans. Advising is the last component of the EIB's operations. As a large investor, the EIB naturally becomes a center of expertise on financing, and therefore offers its advice to other stakeholders. This can for example be in terms of advising on how to carry out projects and find alternative financing, or policy advice on how to create a regulatory framework at country scale to incite investments[4].

The EIB has over the years been gradually expanding its focus on green finance, amounting to a current commitment of 25% of financing towards low – carbon and climate resilient growth. 10% of the EIB's lending takes place outside the EU, of which 35% is committed to green financing.

---

[1] EIB. Part of the EU family[EB/OL]. http:// www. eib. org/about/eu – family/index. htm
[2] EIB. Overview of EIB[EB/OL]. http://www. eib. org/investor_relations/overview/index. htm
[3] EIB. Overview of EIB[EB/OL]. http://www. eib. org/investor_relations/overview/index. htm
[4] EIB. The European Investment Bank at a glance[EB/OL]. http://www. eib. org/infocentre/publications/all/the – eib – at – a – glance. htm

Furthermore, the EIB has been a pioneer in green finance, issuing the world's first green bond in 2007. As a clear indicator of its approach to green finance, the EIB's Climate Strategy states: "*To play a leading role, amongst financial institutions, in mobilizing the finance needed to achieve the worldwide commitment to keep global warming below 2℃ and to adapt to the impacts of climate change*". ①

3. Cooperation Potential of AIIB and EIB

The cooperation potential between the AIIB and the EIB is as illustrated in figure 5.2 below. Square indicates a clear possibility for cooperation, round indicates that no strong factors either enhance or inhibit cooperation within the component, while triangle has clear factors disallowing cooperation.

Figure 5 – 2 Cooperation Framework between AIIB and EIB

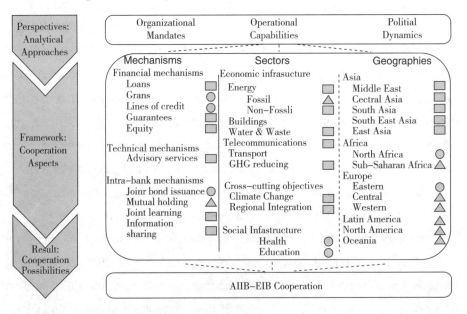

Analysis of the **mandating documents** of the AIIB and EIB determines that the mandate overlap is substantial. While loans, guarantees and

---

① EIB. The European Investment Bank at a glance [EB/OL]. http://www.eib.org/infocentre/publications/all/the-eib-at-a-glance.htm

V. Cooperation of Multilateral Development Banks

knowledge sharing are clear cooperation priorities of both institutions, there is basically no mechanism, sector, or geography that can be completely excluded based on their mandates. The analysis of **operational capacities** determines that the EIB's technical authority as a global and experienced MDB can be well complemented with the AIIB's local legitimacy as an Asian and developing country led MDB. When it comes to the **political dynamics**, the analysis finds that the EU prioritizes energy and transport infrastructure, preferable in the proximity to the EU. It is further found that it is in the interest of China that the AIIB cooperates to the highest extend possible with other MDBs, including the EIB, as this decreases project risks and highlights the AIIB's establishment on par with existing MDBs. This includes all mechanisms, most sectors as highlighted in the Belt and Road Initiative, and on a broad geographic scale with slight preference to countries nearest to China. It is found that Asian developing countries as members of the AIIB and as primary project recipient countries are enthusiastic to attract most forms of infrastructure financing and are consequently in support of MDB cooperation. Lastly, non – AIIB members, including the US and Japan, will not be able to pose any substantial limitations to AIIB – EIB cooperation in terms of any component of the cooperation framework due to the already substantial support globally and regionally pushing the AIIB past the tipping point on legitimacy. US and Japanese opposition might even play to the AIIB's advantage as it can operate independently of their influence. In total, it is clear from the above findings that both institutions can benefit from cooperation at an equal footing.

4. The cooperation potential on climate finance and green finance

Based on the above analysis on how the AIIB and EIB can cooperate in general, it is possible to go in depth with concrete possibilities within green finance. This can be done by specifically approaching green finance aspects within mechanisms, sectors and geographies. In terms of mechanisms, the analysis of organizational mandates, operational capabilities, and political dynamics has shown that loans, guarantees and knowledge sharing are cooperation priorities for both parties. This remains fundamentally true for green finance. As the EIB has been an innovator within green finance, their technical expertise in terms of financial tools and project types can be of great benefit to the AIIB. Of particular interest is the possibility of cooperation towards the AIIB to issuing green bonds. The global green bond market has grown exponentially over the last years, and MDBs have been important initiators and contributors to this trend. Highlighting this current trend, the ADB issued USD 1.25 billion of green bonds on 2 August 2017①.

In this regard, the EIB can assist the AIIB in launching their first green bonds, based on lessons from the EIB's own experiences. This could, in its simplest form, be in terms of intra – bank knowledge transfer. Alternatively, a more ambitious innovative approach is a joint green bond issuance, for example as an "Asian Green Infrastructure Bond". In this way the AIIB can rely on the EIB's expertise to launch itself into the green bond market, in order to subsequently issue green bonds itself. However, the AIIB has yet to begin raising capital on global financial markets. With a total capital endowment of USD 100 billion, taking on liabilities is not yet a critical necessity. Still, with global capital markets' currently great appetite for green fixed income securities, the AIIB could use the occasion to establish themselves in the green bond market. The advantage to the EIB is to both use the joint issuance to increase its currently scare involvement in Asia as well as expand its non –EU green lending to reach its commitment to 35%. Cooperation on green bonds would simultaneously send a strong message on EU –Asia joint commitment to green finance.

When it comes to sectors, the critical aspects to consider are the two MDBs different definitions of green. While they both prioritize renewables and energy efficiency investments, they differ in regards to fossil fuels. In

---

① ADB. ADB Sells Dual –Tranche $ 750 Million 5 –Year and $ 500 Million 10 –Year[EB/OL]. https://www.adb.org/news/adb-dual-tranche-global-green-bonds-spur-climate-financing

V. Cooperation of Multilateral Development Banks

their investments both inside and outside the EU, the EIB rules out projects whose expected emissions are outside national targets, accepts all energy efficiency projects, and only includes transport project with a high socio – economic return[1]. Such concerns can for example be the construction of a highway that would increase emissions from cars and trucks, and consequently the EIB would prefer to finance a railroad or light – rail public transport as this serves the same purpose at a lower emissions rate.

The AIIB, however, includes some fossil fuels under specific circumstances. As of the AIIB's energy strategy, the key phrasing in determining whether circumstances allow for such investments includes being "… demonstrably compatible with a country's transition …", "… use commercially available least – carbon technology.", "Carbon efficient oil – and coal – fired power plants would be considered if they replace existing less efficient capacity or are essential to the reliability and integrity of the system, …"[2]. This difference between AIIB and EIB definitions, is a common occurrence between developed countries, where green is climate change focused and in most cases excludes all fossil fuels, and developing countries, where green goes beyond climate change and includes the least polluting commercially available alternative. Consequently, green finance cooperation is limited to being within the EIB's narrower scope.

Another aspect to consider is that many technical components of green investments belong to companies originating in developed countries, in particular the EU. In representing the EU, the EIB may consequently have an incentive to cooperate with the AIIB in green finance to the highest extent possible, in order to provide European businesses with a gateway to many Asian countries. Compared with other sectorial areas of AIIB –EIB cooperation, green investments usually have a greater technological component, inciting the

---

[1] EIB. Part of the EU family[E/OL]. http://www.eib.org/about/eu –family/index.htm
[2] AIIB. Energy Sector Strategy: Sustainable Energy for Asia[EB/OL]. https://www.aiib.org/en/policies –strategies/strategies/sustainable –energy –asia/index.html

EIB to promote such cooperation over other sectorial categories. On the other hand, green investments often carry a higher risk, due to for example utilizing less established technologies or a more frequently changing political environment. As shown above, since the AIIB is taking a conservative approach to establishing a solid balance sheet, taking on risky investments may be less attractive. Still, in cooperation with the EIB such projects may be less risky than taking them on singlehandedly.

Regarding the geography of green finance cooperation, while the EIB is globally involved, the AIIB is focused on Asia. However, the AIIB has already indicated an interest in other developing countries such as South – Africa, Egypt and Brazil[①]. Countries with particular potential for AIIB – EIB green finance cooperation include Asian countries that are either proactive within green energy or where the energy needs grow very fast. These in particular include Pakistan, India and Indonesia. Further, with limited EIB involvement in this area, Asian countries closer to Europe may also be particularly relevant, such as Azerbaijan and Turkey.

---

① Institute of Development Studies. The Asian Infrastructure Investment Bank: What Can It Learn From, and Perhaps Teach To, the Multilateral Development Banks? [R/OL]. http://www.ids.ac.uk/publication/the – asian – infrastructure – investment – bank – what – can – it – learn – from – and – perhaps – teach – to – the – multilateral – development – banks

VI. Policy Suggestions

(1) **Stick to the Paris Agreement and Promote the Establishment of Global Integrated Climate Public Goods Supply Mechanism**

In the background of the United States' significant change on climate policy resulting from a government shift, China's role to shoulder the responsibility of climate governance is more important than ever.

The root cause of the existing fact that climate financing relies heavily on ODA is that a mature global climate public goods supply concept and financing system has not yet been established. The prerequisite of increasing the quantity, efficiency and effect of climate finance is to form an independent financing mechanism and corresponding financing instruments for mitigation and adaptation. Particularly, it is required to reiterate the core role of global public finance and strengthen the supply and effective use of grant resources, so as to mobilize, lead and promote private sector to solve the problem of insufficient supply of public climate goods. Meanwhile, it is also important to avoid providing global public goods at the expense of national development. Core measures and key fields include:

1. Give full play to the global influence of multilateral platforms and institutions participated and led by China. Promote the international community and multilateral development partnerships to reshape the dual-track mechanism for international cooperation, and establish parallel and coordinated climate public goods supply mechanisms, financing system and toolbox with ODA.

2. Advocate that the governments of the conference of the parties

contribute to climate finance under the principle of "common but differentiated responsibilities". In order to fully fulfill the responsibility, they need to allocate specific quota to relevant ministries, for instance, Environment, Energy, Finance, Health or Trade.

3. Establish more comprehensive international public goods financing institutions to undertake the responsibilities similar to that of OECD Development Assistance Committee.

4. Under the dual-track model of ODA and climate finance, promote the innovation and development of multilateral institutions' rules and regulations, departmental arrangements, funding arrangements and financing tools.

(Ⅱ) Promote the Role and Innovation of Emerging Multilateral Institutions in Addressing Climate Change

The new multilateral institutions led by China cumulatively symbolize China's "soft power" in global climate governance. Based on original aid, investment and trade activities, the role and function of these multilateral institutions further require matching with the leading role of the supply framework of China's global public goods. We recommend that a unified top-level design should be introduced into future development in order to promote innovation of those new multilateral institutions, including governance structure and financing instruments. The main recommendations are as follows:

1. Facilitate infrastructure investment which is compliant with climate resilience goals. The countries along the Belt and Road and in the southern hemisphere are the regions with relatively high global climate risks, and their economy and society are the most vulnerable to climate change. The emerging multilateral institutions should highlight the knowledge and capacity-building, standard development, business mode innovation and mobilize private investment in the field of low-carbon and climate-resilience infrastructures, establish special expert committees and departments, set up corresponding performance assessment standards, and enhance capital and personnel input.

VI. Policy Suggestions

2. At the macro level, countries' mission to meet the needs of residents' life quality and living environment is related to the sustainable development goals. Therefore, the emerging multilateral institutions should take into full consideration the climate goals and policies of the recipient country, refer to the Paris Agreement, and guarantee the compliance of project goals with the shared expectation of all parties considering the "Intended Nationally Determined Contributions" proposed by each country. At the operation level, multilateral institutions may utilize the innovative trading structure, low – risk technologies and local market expertise to mobilize private investment into low – carbon infrastructure construction and other green projects.

3. Stress the cooperation and complementation with traditional multilateral institutions. Though emerging multilateral institutions and traditional multilateral institutions have competition relationships, the cooperation and complementation should be highlighted. Developing countries still face huge financial gaps in infrastructure development, requiring the close cooperation with all multilateral institutions to fill the gap. Emerging economies are still actively supporting the development and reform of traditional multilateral institutions, and they will also seek more opportunities in emerging institutions. In the future, traditional and emerging multilateral institutions can carry out cooperation through joint investment in particular projects and knowledge sharing, thus establishing a new international development financing system to better meet the needs of the developing countries.

4. Encourage mixed financing. In developing countries, there is a relatively large insufficiency in public finance, while the inflow of private finance is often hindered, so it is necessary to scale up both public and private investment through emerging multilateral institutions. Meanwhile, the mixed financing of multilateral and local development institutions is also imperative. Local development banks have a better understanding on local

projects and local regulations, while international or regional organizations have a deeper understanding on the best engineering, design and financial practices. Therefore, mixed financing can expand the financing size on one hand, and can improve the use efficiency of capital thus building a well-functioning system on the other hand.

5. Stress local factors. Some traditional multilateral institutions used to deem developed countries as the core participants in international development, and tried to take the model of developed countries as a template to help developing countries move forward. However, developing countries have the capacity to play a more important role in the international development, and each developing country has its own special national conditions, so there are few template paths that could be followed. It is the original intention of emerging multilateral institutions to address the local issues of developing countries, and this concept must be integrated into climate financing. Exposed to climate change, developing countries are more vulnerable than developed countries, but also have their owner unique power. Developing countries should have comprehensive communication and mutual learning, and find out problems and propose solutions according to actual national situations, thus integrating resources and carrying out business more effectively.

(Ⅲ) **Starting from South – South Cooperation on Climate Change to Strengthen China's Soft Power of Climate Financing**

The establishment of China's Climate Change South – South Cooperation Fund promotes the further development of South – South cooperation on climate change. However, currently China's foreign aid, foreign investment and trade still lack long-term integrated planning and have not yet formed an independent international development assistance system and management structure. Drawing on China's experience in international cooperation in the areas of climate change, regional cooperation and green finance, and the advantages on policy practice, methodology and technology of energy

VI. Policy Suggestions

efficiency and emission reduction, we propose the following recommendations:

1. Establish the South-South cooperation management system and result evaluation mechanism. Foreign aiding countries represented by developed countries have set up different levels of management framework system. From the perspective of more successful experience, such management framework mainly includes four aspects, i.e. legal and policy foundation, organization structure, management mode and cooperation mode. The climate change South-South cooperation is a relatively new field with specific focus in China's foreign aiding, and needs the establishment of an effective management system to guarantee the successful implementation of the cooperation. In addition, to ensure the implementation effect, it is required to establish and improve the result evaluation mechanism and adopt the mode of combining macro assessment and micro assessment. Macro assessment aims to summarize the general effects of the South-South cooperation within a certain period, while the micro assessment is carried out for specific projects. A great performance assessment is beneficial to project design and management, experience summarization and feedback, and is conducive to the improvement of transparency and public credibility.

2. Establish diversified climate financing mechanism. In September 2015, China announced the contribution of RMB 20 billion to establish "China Climate Change South-South Cooperation Fund" to support other developing countries in addressing climate change. The capital sources of the South-South Cooperation should not be limited to donation, and it is required to attract diversified capital, mobilize professional fund management modes, and increase the use efficiency and effect of the capital. In order to attract diversified capital, it requires the project to be "profitable", certain "green investment" may be considered for South-South Cooperation Fund. Such "green investment" is featured by the joint investment and risk sharing by diversified funding channels including private capital which may get a certain proportion of return or other indirect returns and the fund may include the remaining return into re-assistance

or investment. In addition, Chinese government and the recipient government may also take tax reduction and exemption or other policy incentives to attract private capital. The South – South Cooperation Fund may, as a fund of funds, contribute to other regional climate funds in different ways. While learning global experience, it seeks to cooperate with traditional multilateral financial institutions such as the World Bank and the emerging multilateral financial institutions such as Asian Infrastructure Investment Bank (AIIB), so as to make breakthroughs in the polices and practice of providing regional climate public goods.

3. Strengthen top – level design capability building and explore innovative financing instrument platforms. In recent years, China's climate financing system has been continuously improved, and has accumulated rich development experience. In the future, China may focus on providing the experience of constructing green finance system and capacity building e. g. building a learning network under the framework of the South – South climate change cooperation fund to promote the policy framework design capacity, which can mobilize private sector. In addition, in order to use the aiding more effectively, China may learn international experience, set up innovative financing instrument platforms, and increase the leverage ratio of climate aiding provided to other developing countries. For example, most countries are in great demand for buying sovereign insurance, but their participation is highly affected by the availability of cheap or free insurance. China can develop plans to enhance the South – South national climate resilience, providing subsidies for the countries that could not buy climate insurance due to high premium. In addition, the development of financial markets in many developing countries is still very immature, and each country faces different challenges and obstacles in leveraging private sector. Therefore, it is necessary to carry out survey on local demands to specify the barriers and develop solutions according to the national conditions of each country. Meanwhile, China may help developing countries develop more financial instruments such as catastrophe bonds.

VI. Policy Suggestions

Current successful catastrophe bonds include Mexico's earthquake catastrophe bond, the experience of which could be learned for the promotion of application in extreme weather accidents. In addition, China may also, through the mode of loss sharing, encourage banks in the other developing recipient countries to provide green loans, so as to reduce risk level of climate financing and promote capital flow.

4. Make full use of "One Belt One Road" initiative and increase investment in sustainable infrastructure in the countries along "One Belt One Road". In 2016, China comprehensively implemented "One Belt One Road" initiative, further creating a new win – win platform for south – south cooperation on climate change. In the future, China may utilize the initiative to support other developing countries, particularly the least developed countries and vulnerable countries, to strengthen their capacity in addressing climate change, explore the sustainable development paths with climate resilience, and strengthen the capital support to the sustainable infrastructure construction of the countries. Infrastructure is the foundation and driving force of economic growth, and its interconnectivity is a priority area for the construction of the "One Belt One Road". However, the extensive economic development modes and fragile natural ecological environments of many emerging economies have aggravated the climate vulnerability of such countries. Therefore, when supporting the infrastructure construction of the countries along the "One Belt One Road", China should take into full consideration of the effects of climate change, include the whole life cycle covering infrastructure design, investment and operation into the considerations of climate risk factors, and refer to international best adaptation practice to support the construction of infrastructure with climate resilience. In addition, China should provide substantial support to those countries to build their capacity on addressing climate change. Climate adaptation includes not only specific engineering projects such as infrastructure, but also the strengthening of climate risk management.

5. Consummate the MRV system establishment of climate adaptation finance. Currently, most data about climate adaptation finance is based on international public adaptation finance, and the MRV for private adaptation finance, and national and regional public adaptation finance is not yet available. Most climate adaptation activities are driven by the domestic demands of developing countries, and are the actions at national and regional level, most of which are supported by budget capital.[①] Though some countries have estimated this part of capital, they still make up a small minority, and the methods used are different, lacking comprehensiveness and comparability. In the future, the Monitoring, Reporting and Verification of the adaptation finance of China and the recipient countries should be strengthened.

[①] Schealatek L, Nakhooda S, Watson C. The Green Climate Fund[EB/OL]. 2016[2017-09-10]. https://www.odi.org/sites/odi.org.uk/files/resource-documents/11050.pdf.